PRACTICAL POWER
SYSTEM OPERATION

PRACTICAL POWER SYSTEM OPERATION

EBRAHIM VAAHEDI

IEEE PRESS

POWER
ENGINEERING

Mohamed E. El-Hawary, *Series Editor*

Library of Congress Cataloging-in-Publication Data:

Vaahedi, Ebrahim.
Practical power system operation / Ebrahim Vaahedi.
    pages   cm
  Includes bibliographical references and index.
  ISBN 978-1-118-39402-1 (hardback)
1. Electric power systems–Management.   2. Operations research.   I. Title.
  TK1001.V324 2014
  621.31–dc23
                                                        2013039735

ISBN: 9781118394021

To
my wife Elahe
and
my daughters Kiana and Niki

CONTENTS

FOREWORD

Many of us who went through engineering schools invested a substantial amount of capital in textbooks. All of us would probably agree that we learned a lot of "engineering" from these textbooks, and most of us, myself included, would also tell you that the textbooks that we loved to hate have been on the shelves since we graduated and seldom picked out and dusted off to help us solve a real-world engineering problem. They do prove to be useful from time to time when we use them as examples to tell our children who follow our footsteps to become engineers that we learned that subject many many years ago. There was one textbook, however, that was different from them all—it was NOT one of those well-polished, nicely glued together, and expensive textbook, but a hand-written collection of notes used in one of our Power Engineering Courses at the University of British Columbia. It had all the fundamental theories that a Power Engineer would need to master. That "book" followed me through my engineer-in-training rotations and was my most frequently used technical reference in my engineering practice, from system planning to system operations. I have often wondered why there aren't more textbooks like this one.

Dr. Vaahedi's new book *Practical Power System Operation* is a perfect answer to my quest. This textbook is "classical electrical engineering theories meet everyday power system operation." In an easy to follow manner, each chapter starts with a description of simple, practical, and real-life tasks in various facets of power system operations, followed by business processes allowing a glimpse into how these day-to-day tasks may align with the larger business picture, then gets into some hard-core maths and formulas and computational methods to drive home the message that every single task that power system professionals—operators and engineers—perform today in planning and operating this most complex machinery called the power grid is built on the fundamental engineering concepts invented by engineers and scientists generations ahead of us, and that we must continue to improve through innovation for the benefit of the users of our product and for the betterment of mankind.

Practical Power System Operation also offers a great historical view of the evolution of power system operations: from the master–slave analog control systems to the digitation of control technology, from randomly select data points to 4-second data sampling of the grid to microsecond sampling of the synchrophasor systems, from off-line studies with conservative operational method to real-time assessment of the system allowing precise system posturing and maximization of the utilization of assets. It describes the traditional and present day challenges in power system operations and offers a glimpse of the technology ahead. The readers will find this one of the most complete and practical textbooks on power system operation.

It is no surprise to me that Dr. Vaahedi was able to achieve this amazing accomplishment. He is always passionate about bringing innovative solutions to day-to-day operation of the power system. His academic research background and understanding of emerging technology and challenges in adopting the state of the art uniquely enables him to develop visionary and sound technology strategy and practical implementation for our control center organization. He has been instrumental in deploying cutting-edge technology solution in our real-time system operations world for many years. The new textbook is a great testament of his accomplishments. Congratulations Ebrahim!

MARTIN HUANG
Vice President
Grid Operations, BC Hydro

PREFACE

About a decade ago, I started teaching courses on "Decision Support Tools in Power System Operation" at the University of British Columbia. Through the years working closely with system operators at BC Hydro, I came to the realization that a technology course provides a partial view of what actually happens at a control center in operating a power system. To provide a complete picture of power system operation, we need to establish the operator's functions, the processes used to deliver the functions, and the technology solutions that enable the processes. A good metaphor for power system operation is the operation of a car. It is impossible to provide a good picture of the operation of a car by only focussing on the details of its technologies without discussing the functions the car is meant to accomplish or the processes (driving guidelines and rules) deployed to operate the car and deliver the functionality. After looking around, I realized that there is actually no single book in power system operation that provides a complete picture. I had a eureka moment about two years ago when I took the challenge to write a book that combines the practitioner's view of system operation with the details of technology solutions.

The journey to write this book was very difficult. It would not have been possible without the support of many individuals and organizations that I am indebted to. First, I would like to thank Mr. Martin Huang, Vice President of BC Hydro Grid Operation, who encouraged me to write the book and provided insightful comments to make the book one of the elements of operator's training program at BC Hydro. I am also indebted to my employer, BC Hydro, for allowing me to use pictures from BC Hydro's control center in the book. I would also like to thank a number of colleagues at BC Hydro: Mr. Tohru Harada for critically reviewing the book; Mr. Ska-Hiish Manuel for developing examples for the book; Dr. Wenyuan Li for his valuable encouragement and advice; and Mr. Brett Hallborg, the Operations Training Manager at BC Hydro, for commenting on the course contents.

I would also like to thank Alstom Grid for their generosity and their commitment to education and training, giving me the permission to use screenshots of their Energy Management Systems and Distribution Management Systems to better reinforce the concepts discussed in the book. A special thanks goes to my friend Mr. Ali Sadjadpour from Alstom Grid who facilitated the approval.

Finally, I am grateful to a number of my professional colleagues who provided insightful comments on the book: Dr. Antonio Conejo, Chair of IEEE Power System Operations Committee and Professor at Universidad de Castilla–La Mancha, Spain; Dr. Mohammad Shahidehpour, Bodine Chair Professor at Ilinois Institute of Technology; and Dr. David Sun, Chief Scientist at Alstom Grid.

EBRAHIM VAAHEDI

GENERAL INTRODUCTION

This book is the only book approaching power system operation in a holistic manner focusing on operators' needs, the processes required to fulfil those needs, as well as enabling technologies to facilitate the processes. It has been written as a textbook for undergraduate and postgraduate students in the power system area as well as professionals in electric utilities, independent system operators (ISOs), consulting companies, electricity regulators, and all other entities dealing with system operation.

The book is composed of 12 chapters and an appendix. The first eight chapters of the book focus on operators' needs, processes, and the enabling technology solutions. Chapters 9, 10, and 11 provide a complete description of the control centers, energy management systems, and distribution management systems. Finally, Chapter 12 covers the evolving and the state-of-the-art solutions in power system operation. The appendix deals with fundamental theoretical concepts to reinforce the understanding of each chapter.

The content of this book has been taught in the following permanent undergraduate/postgraduate courses at the University of British Columbia in the past decade:

- EECE499: Decision Support Tools in Power System Operation
- EECE498: Application of Optimization in Power System
- EECE553: Advanced Power System Analysis

Classroom adoption potential extends beyond these courses. Other adoption possibilities either in the form of university courses or continuing education courses include:

- Modern Power System Operation
- Power System Control Centers
- Energy Management Systems and Distribution Management Systems
- Computer Applications in Power System Operations

All of the universities around the world that teach power system courses at the undergraduate or postgraduate level can adopt this book.

This book can be used by professionals in electric utilities, ISOs, reliability coordinators, consulting companies, and all other entities involved in power system operation. It can also be used as a reference textbook for operator training and certification.

CHAPTER 1

INTRODUCTION

1.1 OVERVIEW OF POWER SYSTEM OPERATION

The main objectives of power system operation are safety, reliability, and efficiency. System operation has always been regarded as a critical function in utilities around the world because it can significantly change the utility's bottom line. System operation affects people's safety, impacts system reliability, and influences operational costs associated with the deployment of transmission and generation resources.

The electricity deregulation in the last decade created a new landscape for the energy industry. This change coupled with the potential for increasing penetration of large amounts of integrated and variable generation and the move toward smart grid, including advancing generation, transmission, and distribution technologies as well as customer enablement technologies continue to increase the complexity of power system operation.

In power system operation, there are three main actors:

- **Operator**, who is responsible to execute different functions
- **Process**, which provides a detailed description of how a function should be executed and
- **Technology**, which enables and facilitates the process

In simple terms, power system operation involves establishing a picture of the prevailing system operating condition by measuring different system signals such as flows and voltages. Using this picture, the operator executes different operation functions resorting to two means of

- The process book or the operating order book which serves as the reference book detailing the actions that operators need to take under different conditions
- The technology or decision support tools which enable and facilitate the operating orders

This exercise will eventually result in some automated control actions directly taken by the system or manual control actions taken by the system operator as shown in Figure 1.1. A good metaphor for this is operating or driving a car. The operator examines the driving condition by watching the surroundings of the car. He compliments this information with the information from the car dashboard, which provides the

Practical Power System Operation, First Edition. Ebrahim Vaahedi.
© 2014 The Institute of Electrical and Electronics Engineers, Inc. Published 2014 by John Wiley & Sons, Inc.

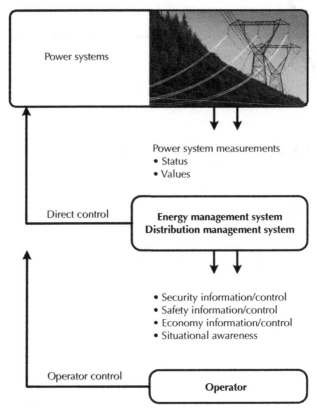

Figure 1.1 Power system operation function.

prevailing operating condition of the vehicle such as speed, engine's temperature, and gas content. The operator then responds to the prevailing condition based on the rules specified in the driving rule book using the technologies enabling the rule book actions such as the steering and brake systems. Under some conditions, the car can take an automated control action directly such as triggering airbags under emergencies using the prevailing condition without operators' interaction.

1.2 OPERATOR

Operators are responsible to manage system operation with the objectives of safety, reliability, and operational efficiency. To achieve these objectives, they need to perform a number of functions and their associated tasks. Some of the major functions that operators need to execute include

- Transmission monitoring
- Transmission limit assessment
- Transmission voltage management

- Transmission congestion management
- Transmission outage management
- Load shedding management
- Generation load balance
- Generation operation
- Generation reserve management
- Generation outage management
- Interchange communication
- Interchange monitoring
- Interchange congestion management
- Interchange operation
- Emergency generation capacity management
- Emergency load shedding management
- Emergency transmission management
- Restoration

North American Electric Reliability Council (NERC) has published a generic task list [1] detailing all the functions that system operators need to execute to manage system operation. The task list includes about 400 tasks divided under the following broad categories:

- General control center operations
- Transmission
- Generation
- Interchange
- Restoration

1.3 PROCESS

To fulfill their functions, operators need to develop business processes for each function. A business process is a structured set of linked activities designed to produce a product or service. One of the most significant people in the eighteenth century to describe processes was Adam Smith [2], who described the production of a pin in the following way:

> One man draws out the wire, another straights it, a third cuts it, a fourth points it, a fifth grinds it at the top for receiving the head: to make the head requires two or three distinct operations: to put it on is a particular business, to whiten the pins is another... and the important business of making a pin is, in this manner, divided into about eighteen distinct operations, which in some manufactories are all performed by distinct hands, though in others the same man will sometime perform two or three of them.

Starting from Chapter 2, the process corresponding to each operator critical function will be examined.

1.4 TECHNOLOGY

To enable and facilitate business processes, technology solutions are required. For example, to achieve the process of pin production with minimum manual work, a number of technology solutions need to be deployed. Similarly, for power system operation, each process may need one or a number of technology solutions for process enablement.

1.5 POWER SYSTEM OPERATION CRITERIA

NERC was established following the blackout of Northeast USA with the mandate of developing operation and planning reliability standards. These standards are the planning and operating rules that electric utilities follow to ensure an acceptable level of reliability. While NERC's standards were historically voluntary, in 2007, the Federal Energy Regulatory Commission (FERC) made them mandatory. This gave NERC another mandate to enforce compliance with NERC Reliability Standards, which it achieves through a rigorous program of monitoring, audits, and investigations and the imposition of financial penalties and other enforcement actions for noncompliance. NERC also manages the program that certifies system operators, ensuring they have the required knowledge and skills to perform their functions.

NERC defines reliability as the combination of transmission adequacy and transmission security where:

- Transmission adequacy is defined as "The ability of the electric system to supply the aggregate demand and energy requirements of their customers at all times, taking into account scheduled and reasonably expected unscheduled outages of system elements/components."
- Transmission security is defined as "The ability of the bulk electric system to withstand sudden disturbances such as electric short circuits, unanticipated loss of system components or switching operations."

For reliable power system operation, a power system must remain intact and be capable of withstanding a variety of disturbances. Since it is impractical to consider all contingencies, instead only a set of probable contingencies are considered. These contingencies, which are called $(N - 1)$, are the ones that consider the loss of only one element of an N-component grid. It is recognized that more severe contingencies than $(N - 1)$ can occur for which a reliable power system operation may not be achieved. For these conditions, measures should be taken to minimize the impact.

NERC standards [3] define the system performance requirement for the normal system when the system has not suffered any contingencies and the conditions after the contingencies. Performance requirement could include voltage profiles, allowed voltage dip after a contingency, system frequency damping, etc. An example of

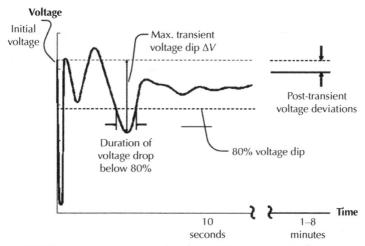

Figure 1.2   NERC voltage performance requirement following a contingency.

voltage profile performance requirement after a contingency is given in Figure 1.2. This figure shows the limits imposed on the maximum transient voltage dip, the time duration of voltage dip exceeding the imposed limit (e.g., 20%), and the post-transient voltage deviation.

1.6   OUTLINE OF THE BOOK

The book is composed of 12 chapters and one appendix. It describes power system operation from the viewpoint of an operator's functions and needs. This practical approach provides the readers with a deep appreciation for power system operation objectives, processes established to achieve those objectives, and the technology needed to enable the processes.

The first eight chapters of the book cover an operator's functions and his needs to fulfill his functions. This is followed by providing the established process details to perform the functions as well as the required technology solutions enabling the processes. Starting from Chapter 2, each chapter is dedicated to one operator's critical function, its process, and the enabling technology. Chapter 9 discusses modern power system operation control centers' requirements and design features. Chapters 10 and 11 provide complete descriptions of energy management systems and distribution management systems. Each chapter provides an ensemble of technology solutions playing together for transmission and distribution operation respectively. Chapter 12 deals with evolving technology solutions for system operation. Finally, Appendix A deals with fundamental theoretical concepts to reinforce the understanding of the chapters.

CHAPTER 2

POWER SYSTEM MONITORING

2.1   OPERATOR FUNCTION IN POWER SYSTEM MONITORING

Power system monitoring is the most fundamental function of a system operator. Operators need to examine the prevailing system condition to establish whether the system is operating within acceptable thresholds. The ideal solution for the operator is to have full visibility by measuring system attributes such as voltages and flows. This, however, has not been the case due to the significant cost associated with vast numbers of measurements rendering the control centers with only a selective number of critical measurements. To fulfill the system monitoring function, the key question the operator is tasked to establish is "is the system OK now?".

2.2   PROCESS FOR POWER SYSTEM MONITORING

The process of monitoring involves the examination of power system conditions. Operators need to examine the prevailing system conditions to establish whether the system is operating within acceptable physical and operational limits. Examples of monitoring parameters include:

- Substation voltages
- Transmission power flows
- Generator active and reactive power generation
- Total system load
- Interchange schedules
- System frequency
- Status of circuit breakers and switches

Operators need to measure these system parameters to be able to monitor the system conditions and establish whether they need to take any further actions to remove potential violations.

A great metaphor for this is the operation of a car. As the car operator drives the car, he needs to observe all the prevailing conditions such as the engine temperature and the car speed using an easy-to-understand human interface called the dashboard.

Practical Power System Operation, First Edition. Ebrahim Vaahedi.
© 2014 The Institute of Electrical and Electronics Engineers, Inc. Published 2014 by John Wiley & Sons, Inc.

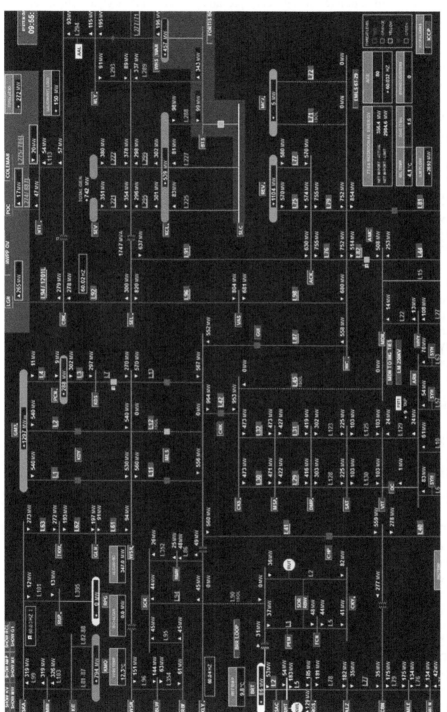

Figure 2.1 System overview diagram providing high-voltage transmission flows, generation and other system information.

The dashboard in turn receives the car's prevailing condition through an infrastructure that measures and transports the information to the dashboard for operator viewing. This infrastructure consists of a number of measurement sensors wired to the dashboard.

A typical example of the system overview needed by system operators for monitoring the system is given in Figure 2.1. This figure shows the high-voltage system including generation at different substations and power flows on the transmission system. The statuses of circuit breakers and switches are also reflected in this figure, indicating whether they are closed or open. The operator is able to zoom on a specific substation or transmission equipment, examine power flows, and open or close circuit breakers or switches.

2.3 TECHNOLOGY FOR POWER SYSTEM MONITORING

The decision system required for power system monitoring includes the infrastructure to measure signals such as voltages and reactive and active power flows, status of switches, and transferring them to the control center. This is followed by using a method to filter the information and recreate the missing information that has not been measured. As shown in Figure 2.2, the infrastructure used for measuring and transporting the information to the control is called Supervisory Control and Data Acquisition (SCADA) system while the decision support tool that filters and recreates the missing information is called "State Estimation."

2.3.1 The Role of System Control and Data Acquisition (SCADA)

SCADA performs the two main functions of data acquisition and supervisory control. SCADA's function is composed of three parts:

1. Data acquisition and switching enablement in the field
2. Communication system
3. Supervisory control

The Data Acquisition and Switching enablement in the field have been conventionally performed using a device called remote terminal unit (RTU). An RTU monitors the field digital and analog parameters and transmits data to the central monitoring station. An RTU can be interfaced with the central station with different communication media (usually serial (RS232, RS485, RS422) or Ethernet). An RTU can support standard protocols (Modbus, IEC 60870-5-101/103/104, DNP3, IEC 60870-6-ICCP, IEC 61850, etc.) to interface to any third-party software. In some control applications, RTUs drive a digital output board to switch power on and off to devices in the field. This is done by the digital output board triggering the relay, which closes the high-current contacts. RTUs are placed in the substations and collect the equipment status and system parameters such as active and reactive power, voltages, etc. The supervisory control system scans all the RTUs on a predefined scan period (e.g., 2 seconds) and collects the information collected by RTUs. The supervisory control system has also

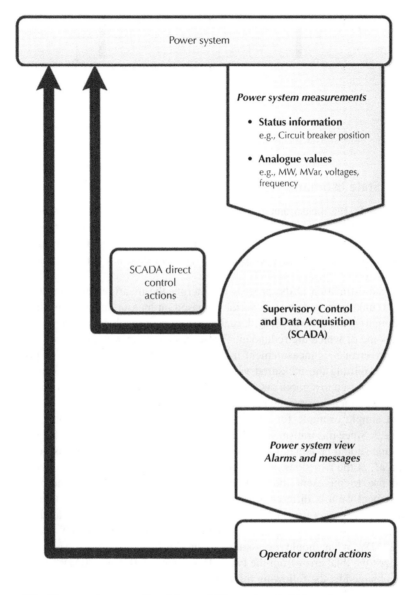

Figure 2.2   Power system operation data acquisition and control hierarchy.

the ability to execute operators' switching requests. These requests are sent to RTUs, which in turn execute the switching by triggering relays associated with switches.

As an example, let us say that after scanning of the RTU information and presenting it to the operator, the operator observes that the voltage at a substation is low and it would be best to switch in a capacitor to increase the voltage at the bus to bring the voltage to an acceptable level. He then proceeds by remotely switching in the capacitor through the action of an RTU, which drives the opening and closing of the capacitor switch.

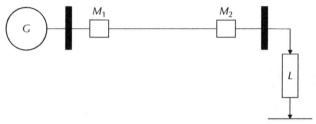

Figure 2.3   Flow measurements on the two sides of a lossless line.

2.3.2   State Estimation

Unfortunately, the measurement devices do not provide an errorless measurement. Normally, each measuring device introduces some random error characterized by its standard deviation. The lower the standard deviation, the less random measurement error the device introduces. Therefore, a filtering mechanism is required to smooth out the measurements.

State estimation is the process of filtering the measurements and recreating values to unknown system state variables based on measurements from the system according to some criteria [4–8]. Usually, the process involves inaccurate measurements, some of which are redundant. These inaccurate measurements are primarily caused by erroneous measurement transducers and from communication problems while transmitting the measured value back to the operation control center. State estimation is used to smooth out random errors in measurements, detect and identify large measurement errors, and recreate unavailable ones.

A simple example for state estimation is a lossless transmission shown in Figure 2.3. Since this transmission line is lossless, the power flow from the two sides of the line should be equal. Let us assume that we have two flow measurements of M_1 and M_2 at the two sides of the transmission line. The question is which one we should use to represent the flow on the transmission line. Depending on the measurement we use, different results may emerge for the system. State estimation attempts to estimate a flow that minimizes the error between the estimated flow and the measurements M_1 and M_2. Let us assume that M_1 indicates 100 MW and M_2 indicates 20 MW. What should we assume as the flow? The idea is to use a value that minimizes the total error from the two data samples of 100 and 20 MW. The best fit can be obtained by the following minimization problem:

$$\text{Minimize } F(x) = (x-100)^2 + (x-20)^2 \tag{2.1}$$

To minimize $F(x)$, we make $F'(x)=0$, which gives

$$F'(x) = 2(x-100) + 2(x-20) = 0 \tag{2.2}$$

resulting in

$$2x - 120 = 0 \quad \text{and} \quad x = 60. \tag{2.3}$$

If we know ahead of time that the measurement device of M_1 is more accurate than M_2, then we would need to give more weighting to M_1 proportional to the degree of its accuracy versus M_2.

2.3.3   Least Square Method for State Estimation

Least square is a mathematical method to obtain the state variable values by minimizing the estimation error. The mathematical formulation can be stated as

$$\min J(x) = \sum_{i=1}^{n} \frac{[Z_i^{\text{meas}} - f_i(x)]^2}{\delta_i^2} \tag{2.4}$$

where, $f_i(x)$, function to calculate the value being measured by the ith measurement; δ_i^2, variance for the ith measurement. The variance value provides a confidence level we assign to a specific measurement. A high value reflects a high error whereas a low value indicates a high-accuracy measurement device; $J(x)$, measurement residual; n, number of independent measurements; Z_i^{meas}, ith measured quantity.

It is also assumed that each measurement provides a value that has a random error with a standard deviation of δ_i. Also, it is assumed that the expected value of the error for each measurement device is zero, meaning that the errors are on both sides of the actual value canceling each other when we have a large number of measurements. The meaning of standard deviation is that 99% of the measurements should fall within $\pm 3\delta_i$. The components of the covariance matrix give the covariance (or the square of the standard deviation δ_i) of the corresponding measurement. The covariance matrix introduces the weighting that must be placed on each measurement as a result of its accuracy. Furthermore, it is worth noting that in Equation 2.2 the error in each measurement has been normalized based on its variance, which is a representation of the device accuracy.

Equation 2.4 can be written in matrix form without loss of generality, that is, assuming that $f_i(x)$ remains nonlinear:

$$F(x) = \begin{bmatrix} f_1(x) \\ \vdots \\ \vdots \\ \vdots \\ f_n(x) \end{bmatrix} \qquad Z^{\text{meas}} = \begin{bmatrix} Z_1^{\text{meas}}(x) \\ \vdots \\ \vdots \\ \vdots \\ Z_n^{\text{meas}}(x) \end{bmatrix} \tag{2.5}$$

Using the vectors given in Equation 2.5, Equation 2.4 can be written as

$$\min J(x) = \left[Z^{\text{meas}} - F(x) \right]^{\text{T}} \left[R^{-1} \right] \left[Z^{\text{meas}} - F(x) \right] \tag{2.6}$$

where R, which is called the covariance matrix of the measurement errors, is given by

$$[R] = \begin{bmatrix} \sigma_1^2 & & \\ & \sigma_2^2 & \\ & & \sigma_n^2 \end{bmatrix} \tag{2.7}$$

To obtain the minimum of J, one would need to solve an unconstrained minimization problem of

$$\frac{\partial J}{\partial x} = 0 \tag{2.8}$$

where the value of x obtained at the minimum is the best estimate of x. By taking the derivative of Equation 2.6, the following n-dimensional vector equation is obtained, which needs to be satisfied:

$$H^{\mathrm{T}}(x) \cdot R^{-1} \cdot \left(Z^{\mathrm{meas}} - F(x) \right) = 0 \tag{2.9}$$

where $H(x)$ is the derivative of $F(x)$ or the Jacobian of $F(x)$:

$$H(x) = \begin{bmatrix} \dfrac{\partial F_1(x)}{\partial x_1} & \dfrac{\partial F_1(x)}{\partial x_2} & \cdots \\ \dfrac{\partial F_n(x)}{\partial x_1} & \dfrac{\partial F_n(x)}{\partial x_2} & \cdots \end{bmatrix} \tag{2.10}$$

To solve the nonlinear Equation 2.9, the Newton–Raphson method, which deploys linearization, can be used. This way the following equation should be solved iteratively:

$$X_{k+1} = X_k + \left(H_k^{\mathrm{T}} \cdot R^{-1} \cdot H_k \right)^{-1} H_k^{\mathrm{T}} \cdot R^{-1} \cdot \left(Z^{\mathrm{meas}} - F(X_k) \right) \tag{2.11}$$

For a linear system where $F(X) = H \cdot X$, the following equation is obtained:

$$X = \left(H^{\mathrm{T}} \cdot R^{-1} \cdot H \right)^{-1} H^{\mathrm{T}} \cdot R^{-1} \cdot Z^{\mathrm{meas}} \tag{2.12}$$

2.3.3.1   Linear System Example 1   For the example given in Figure 2.3, let us assume that the standard deviation for M_1 is 0.001 and the standard deviation for M_2 is 0.003. What would be the estimated flow using the least square method?

Solution
Using the least square method, we have $f_1(x) = x$ and $f_2(x) = x$:

$$F(x) = \begin{bmatrix} f_1(x) \\ f_2(x) \end{bmatrix} = \begin{bmatrix} x \\ x \end{bmatrix}$$

$$Z^{\mathrm{meas}} = \begin{bmatrix} M_1 \\ M_2 \end{bmatrix} = \begin{bmatrix} 100 \\ 20 \end{bmatrix}$$

and

$$[R] = \begin{bmatrix} 0.001 & \\ & 0.003 \end{bmatrix}$$

Using the aforementioned vectors in Equation 2.9 gives

$$H^{\mathrm{T}}(x) \cdot R^{-1} \cdot \left(Z^{\mathrm{meas}} - F(x) \right) = 0$$

$$[1 \quad 1] \begin{bmatrix} 0.001 & \\ & 0.003 \end{bmatrix}^{-1} \begin{bmatrix} 100 - x \\ 20 - x \end{bmatrix} = 0$$

Solving for x, we get $x = 80\,\mathrm{MW}$, which is greater than the value obtained in Section 2.3.2.

2.3.3.2 *Linear System Example 2*   Consider an example with the network configuration shown in Figure 2.4:

$$Z^{\mathrm{meas}} = \begin{bmatrix} Z_{12}^{\mathrm{Meas}} \\ Z_{13}^{\mathrm{Meas}} \\ Z_{23}^{\mathrm{Meas}} \end{bmatrix} = \begin{bmatrix} 0.6 \\ 0.7 \\ 0.4 \end{bmatrix}$$

The network has the following equations:

$$M_{12} = f_1(\theta_1, \theta_2) = 4\theta_1 - 4\theta_2 \tag{2.13}$$

$$M_{13} = f_2(\theta_1, \theta_2) = 2\theta_1 \tag{2.14}$$

$$M_{32} = f_3(\theta_1, \theta_2) = -5\theta_2 \tag{2.15}$$

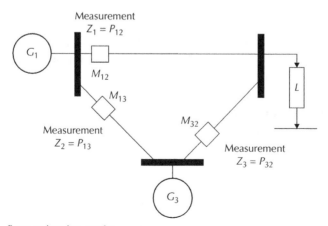

Figure 2.4   State estimation results.

From the aforementioned equations, we obtain H as:

$$H(x) = \begin{bmatrix} \dfrac{\partial f_1}{\partial x_1} & \dfrac{\partial f_1}{\partial x_2} \\[6pt] \dfrac{\partial f_2}{\partial x_1} & \dfrac{\partial f_2}{\partial x_2} \\[6pt] \dfrac{\partial f_3}{\partial x_1} & \dfrac{\partial f_3}{\partial x_2} \end{bmatrix} = \begin{bmatrix} 4 & -4 \\ 2 & 0 \\ 0 & -5 \end{bmatrix} \tag{2.16}$$

Using the covariance matrix $[R]$ given next:

$$[R] = \begin{bmatrix} \sigma^2_{m11} & & \\ & \sigma^2_{m22} & \\ & & \sigma^2_{m33} \end{bmatrix} = \begin{bmatrix} 3 \times 10^{-4} & & \\ & 2 \times 10^{-4} & \\ & & 1 \times 10^{-4} \end{bmatrix} \tag{2.17}$$

the following values are obtained for Θ_1 and Θ_2:

$$\begin{bmatrix} \Theta_1^{est} \\ \Theta_2^{est} \end{bmatrix} = \left[\begin{bmatrix} 4 & 2 & 0 \\ -4 & 0 & -5 \end{bmatrix} \begin{bmatrix} 0.0003 & & \\ & 0.0002 & \\ & & 0.0001 \end{bmatrix}^{-1} \begin{bmatrix} 4 & -4 \\ 2 & 0 \\ 0 & -5 \end{bmatrix} \right]^{-1} \cdot$$

$$\left[\begin{bmatrix} 4 & 2 & 0 \\ -4 & 0 & -5 \end{bmatrix} \begin{bmatrix} 0.0003 & & \\ & 0.0002 & \\ & & 0.0001 \end{bmatrix}^{-1} \begin{bmatrix} 0.6 \\ 0.7 \\ 0.4 \end{bmatrix} \right]$$

$$\begin{bmatrix} \Theta_1^{est} \\ \Theta_2^{est} \end{bmatrix} = \left[\begin{bmatrix} 73,333 & -53,333 \\ -53,333 & 303,333 \end{bmatrix}^{-1} \begin{bmatrix} 15,000 \\ -28,000 \end{bmatrix} \right] = \begin{bmatrix} 0.015756 \\ -0.0646 \end{bmatrix}$$

From these values, the network flows can be calculated using equations 2.13–2.15. Since the variance element σ_{m33} is much smaller than the other two variances, the flow corresponding to σ_{m33} should be closer to its measurement than the flows corresponding to the other two variances respectively.

2.3.3.3 Nonlinear System Example

To demonstrate a nonlinear case, the two-bus system of Figure 2.5 is considered.

Given $V_1 = 1.0$ and that the measured values are:

$$Z_1 = V_2 = 1.02$$

$$Z_2 = P_{12} = 2.0$$

$$Z_3 = Q_{12} = 0.2$$

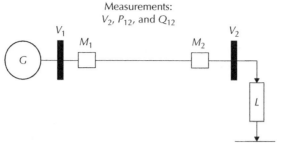

Figure 2.5   The two-bus system.

and that the error characteristics of the measurement devices are given with the following covariance matrix:

$$[R] = \begin{bmatrix} (0.05)^2 & & \\ & (0.1)^2 & \\ & & (0.15)^2 \end{bmatrix} \tag{2.18}$$

find the best estimates of V_2 and δ_2 assuming the following initial conditions:

$$\begin{bmatrix} x_1 \\ x_2 \end{bmatrix} = \begin{bmatrix} \delta_2 \\ V_2 \end{bmatrix} = \begin{bmatrix} 0 \\ 1.02 \end{bmatrix} \tag{2.19}$$

To find the best estimate for the aforementioned values, we first establish the equations relating the measurements to state variables:

$$f_1(\delta_2, V_2) = V_2 \tag{2.20}$$

$$f_2(\delta_2, V_2) = -10 V_2 \sin \delta_2 \tag{2.21}$$

$$f_3(\delta_2, V_2) = 10 - 10 V_2 \cos \delta_2 \tag{2.22}$$

Based on these values, we establish the Jacobian $H(\delta_2, V_2)$ as given next:

$$H(\delta_2, V_2) = \begin{bmatrix} df_1/d\delta_2 & df_1/dV_2 \\ df_2/d\delta_2 & df_2/dV_2 \\ df_3/d\delta_2 & df_3/dV_2 \end{bmatrix} \tag{2.23}$$

$$H(\delta_2, V_2) = \begin{bmatrix} 0 & 1 \\ -10 V_2 \cos \delta_2 & -10 \sin \delta_2 \\ 10 V_2 \sin \delta_2 & -10 \cos \delta_2 \end{bmatrix} \tag{2.24}$$

This results in the following values:

$$z - f(0, 1.02) = \begin{bmatrix} 0 \\ 2.0 \\ 0.2 \end{bmatrix} \quad H_0(0, 1.02) = \begin{bmatrix} 0 & 1 \\ -10.2 & 0 \\ 0 & -10 \end{bmatrix}$$

$$H_0^T R^{-1}(z - f_0) = \begin{bmatrix} 0 & -10.2 & 0 \\ 1 & 0 & -10 \end{bmatrix} \begin{bmatrix} 400 & 0 & 0 \\ 0 & 100 & 0 \\ 0 & 0 & 44.4 \end{bmatrix} \begin{bmatrix} 0 \\ 2.0 \\ 0.4 \end{bmatrix}$$

$$= \begin{bmatrix} 0 & -1020 & 0 \\ 400 & 0 & -444.4 \end{bmatrix} \begin{bmatrix} 0 \\ 2.0 \\ 0.4 \end{bmatrix}$$

$$= \begin{bmatrix} -2040 \\ -177.8 \end{bmatrix}$$

$$\left(H_0^T R^{-1} H_0 \right)^{-1} = \left(\begin{bmatrix} 0 & -10.2 & 0 \\ 1 & 0 & -10 \end{bmatrix} \begin{bmatrix} 400 & 0 & 0 \\ 0 & 100 & 0 \\ 0 & 0 & 44.4 \end{bmatrix} \begin{bmatrix} 0 & 1 \\ -10.2 & 0 \\ 0 & -10 \end{bmatrix} \right)^{-1}$$

$$= \left(\begin{bmatrix} 10,404 & 0 \\ 0 & 4,844.4 \end{bmatrix} \right)^{-1}$$

This gives

$$X^{k+1} = X^k + [H^T_k . R^{-1} . H_k]^{-1} [H^T . R^{-1} . (Z^{\text{means}} - f(X_k)]$$

$$\begin{bmatrix} x_1^1 \\ x_2^1 \end{bmatrix} = \begin{bmatrix} 0 \\ 1.02 \end{bmatrix} + \begin{bmatrix} 10404 & 0 \\ 0 & 4844.4 \end{bmatrix}^{-1} \begin{bmatrix} -2040 \\ -177.8 \end{bmatrix}$$

$$= \begin{bmatrix} -0.196 \\ 0.983 \end{bmatrix}$$

From here, we proceed to the second iteration, and so on.

2.4 BAD DATA IDENTIFICATION

Knowing that 99% of the measurements should fall within 99% within ±3 of the measurement standard deviation, it is possible to find measurements that are grossly wrong. Once the estimation is done, one can calculate the residual for each measurement and divide it by the standard deviation of that measurement to get a normalized value. If this value is greater than 3, then it is highly suspected that the

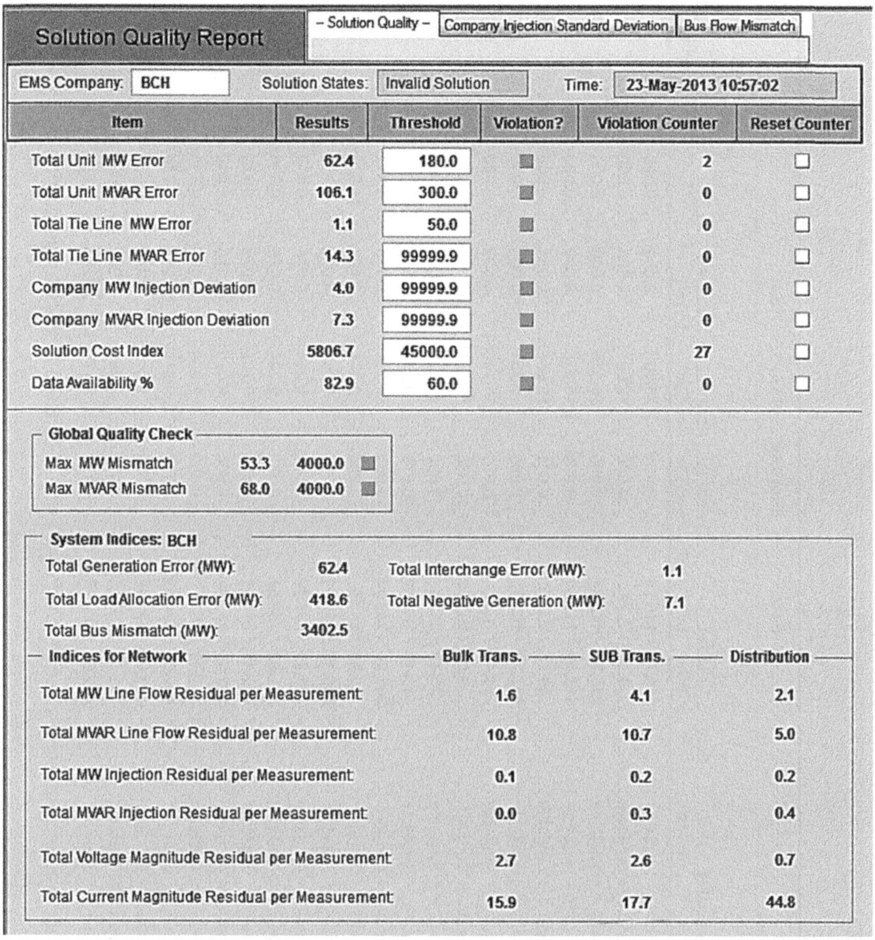

Solution Quality Report	– Solution Quality –	Company Injection Standard Deviation	Bus Flow Mismatch

EMS Company: BCH	Solution States: Invalid Solution	Time: 23-May-2013 10:57:02

Item	Results	Threshold	Violation?	Violation Counter	Reset Counter
Total Unit MW Error	62.4	180.0	▪	2	☐
Total Unit MVAR Error	106.1	300.0	▪	0	☐
Total Tie Line MW Error	1.1	50.0	▪	0	☐
Total Tie Line MVAR Error	14.3	99999.9	▪	0	☐
Company MW Injection Deviation	4.0	99999.9	▪	0	☐
Company MVAR Injection Deviation	7.3	99999.9	▪	0	☐
Solution Cost Index	5806.7	45000.0	▪	27	☐
Data Availability %	82.9	60.0	▪	0	☐

Global Quality Check
Max MW Mismatch	53.3	4000.0 ▪
Max MVAR Mismatch	68.0	4000.0 ▪

System Indices: BCH
Total Generation Error (MW):	62.4	Total Interchange Error (MW):	1.1
Total Load Allocation Error (MW):	418.6	Total Negative Generation (MW):	7.1
Total Bus Mismatch (MW):	3402.5		

Indices for Network	Bulk Trans.	SUB Trans.	Distribution
Total MW Line Flow Residual per Measurement:	1.6	4.1	2.1
Total MVAR Line Flow Residual per Measurement:	10.8	10.7	5.0
Total MW Injection Residual per Measurement:	0.1	0.2	0.2
Total MVAR Injection Residual per Measurement:	0.0	0.3	0.4
Total Voltage Magnitude Residual per Measurement:	2.7	2.6	0.7
Total Current Magnitude Residual per Measurement:	15.9	17.7	44.8

Figure 2.6   A typical state estimation display indicating the quality of the solution.

measurement is bad. By analyzing the residuals, one can start the hypothesis testing process by throwing the measurement with the largest residual and performing the estimation process again. If that fails to yield acceptable estimates, the thrown estimate is put back and the measurement with the largest residual in the new estimate is removed. Meanwhile to establish whether the thrown-out measurement is truly bad, the new value of $J(x)$ is compared with its old value and the change is noted. If the change is negative, then the hypothesis of bad measurement is correct and vice versa. Another measure of the proper performance of state estimators is the value of $J(x)$. It is possible to show statistically the expected value of $J(x)$, and hence one can establish a maximum ceiling that with a high probability $J(x)$ is smaller than. Figure 2.6 shows a typical state estimation display in an energy management system (EMS) indicating the solution quality. Figure 2.7 on the other hand shows a typical list of telemetered measurement signals in an EMS comparing the measured values against the estimated values. Also shown in this display are the suspect measurements.

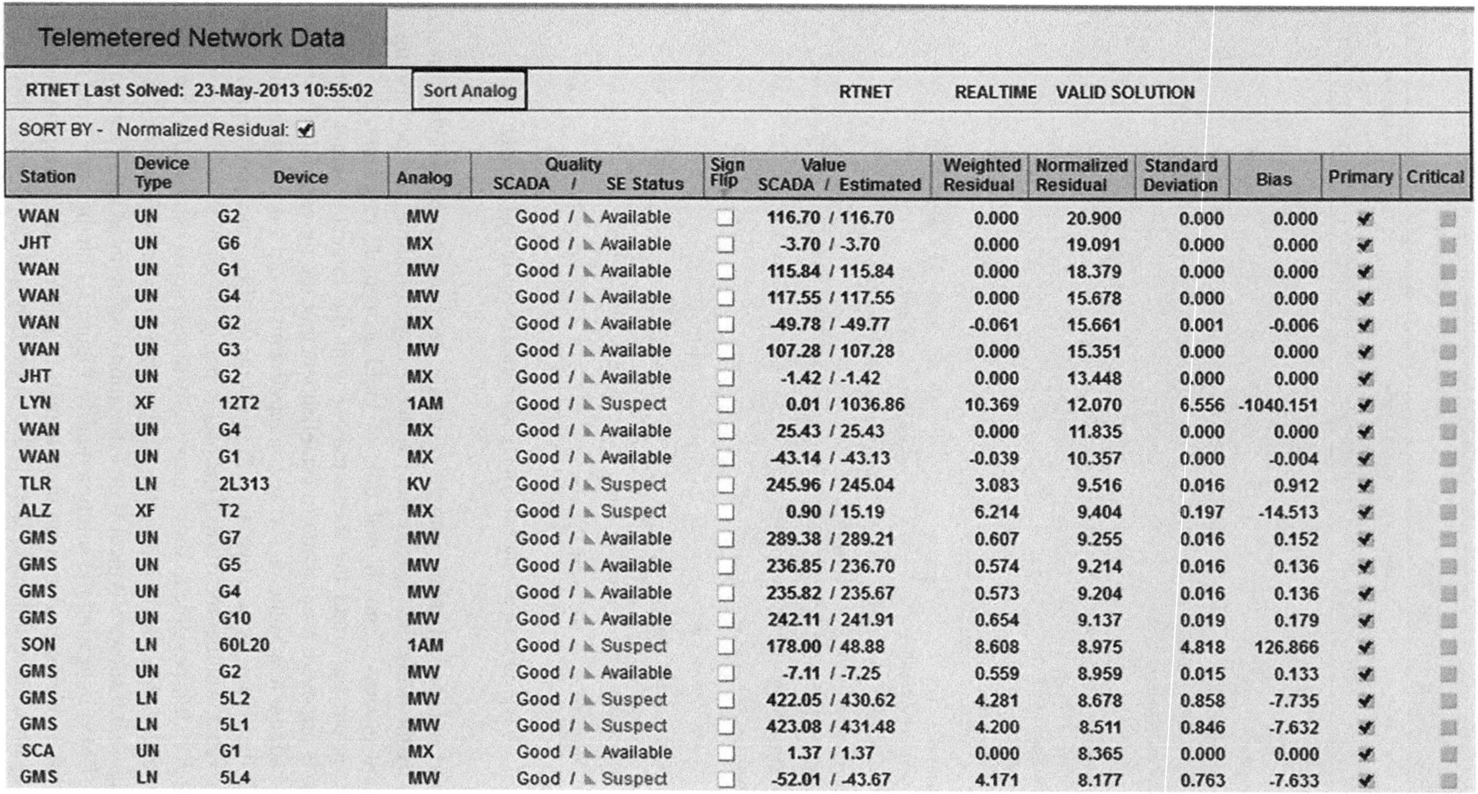

Telemetered Network Data

| RTNET Last Solved: 23-May-2013 10:55:02 | Sort Analog | | RTNET | REALTIME | VALID SOLUTION |

SORT BY - Normalized Residual: ✓

Station	Device Type	Device	Analog	Quality SCADA / SE Status	Sign Flip	Value SCADA / Estimated	Weighted Residual	Normalized Residual	Standard Deviation	Bias	Primary	Critical
WAN	UN	G2	MW	Good / Available		116.70 / 116.70	0.000	20.900	0.000	0.000	✔	
JHT	UN	G6	MX	Good / Available		-3.70 / -3.70	0.000	19.091	0.000	0.000	✔	
WAN	UN	G1	MW	Good / Available		115.84 / 115.84	0.000	18.379	0.000	0.000	✔	
WAN	UN	G4	MW	Good / Available		117.55 / 117.55	0.000	15.678	0.000	0.000	✔	
WAN	UN	G2	MX	Good / Available		-49.78 / -49.77	-0.061	15.661	0.001	-0.006	✔	
WAN	UN	G3	MW	Good / Available		107.28 / 107.28	0.000	15.351	0.000	0.000	✔	
JHT	UN	G2	MX	Good / Available		-1.42 / -1.42	0.000	13.448	0.000	0.000	✔	
LYN	XF	12T2	1AM	Good / Suspect		0.01 / 1036.86	10.369	12.070	6.556	-1040.151	✔	
WAN	UN	G4	MX	Good / Available		25.43 / 25.43	0.000	11.835	0.000	0.000	✔	
WAN	UN	G1	MX	Good / Available		-43.14 / -43.13	-0.039	10.357	0.000	-0.004	✔	
TLR	LN	2L313	KV	Good / Suspect		245.96 / 245.04	3.083	9.516	0.016	0.912	✔	
ALZ	XF	T2	MX	Good / Suspect		0.90 / 15.19	6.214	9.404	0.197	-14.513	✔	
GMS	UN	G7	MW	Good / Available		289.38 / 289.21	0.607	9.255	0.016	0.152	✔	
GMS	UN	G5	MW	Good / Available		236.85 / 236.70	0.574	9.214	0.016	0.136	✔	
GMS	UN	G4	MW	Good / Available		235.82 / 235.67	0.573	9.204	0.016	0.136	✔	
GMS	UN	G10	MW	Good / Available		242.11 / 241.91	0.654	9.137	0.019	0.179	✔	
SON	LN	60L20	1AM	Good / Suspect		178.00 / 48.88	8.608	8.975	4.818	126.866	✔	
GMS	UN	G2	MW	Good / Available		-7.11 / -7.25	0.559	8.959	0.015	0.133	✔	
GMS	LN	5L2	MW	Good / Suspect		422.05 / 430.62	4.281	8.678	0.858	-7.735	✔	
GMS	LN	5L1	MW	Good / Suspect		423.08 / 431.48	4.200	8.511	0.846	-7.632	✔	
SCA	UN	G1	MX	Good / Available		1.37 / 1.37	0.000	8.365	0.000	0.000	✔	
GMS	LN	5L4	MW	Good / Suspect		-52.01 / -43.67	4.171	8.177	0.763	-7.633	✔	

Figure 2.7 A typical EMS display comparing the telemetered measurement values against their estimated values identifying the suspect measurements.

2.5   OBSERVABILITY

In the state estimation, it is necessary to have the number of measurements more than the typical state variable to be estimated. In other words, it is necessary to have redundant information for estimation. Even this condition does not warrantee that all the state variables can be estimated from the measurements because of "observability" phenomena. In summary, observability is the ability of the measurements to observe the state variables. For example, there could be measurements that do not have information about a state variable. Traditionally, there are methods to establish if a system is observable with a set of measurements. Recently, there have been methods developed that not only identify if a system is observable or not, they can identify how much a single measurement observes different state variables and basically comes up with quantitative observability. In other words, we may have a system to be declared observable, but at the same time we can establish that some state variable cannot be strongly observable.

QUESTIONS AND PROBLEMS

2.1. If the operators had perfect measurements from all the system, would there be any need for the state estimation?

2.2. Draw the power system operation data acquisition and control hierarchy diagram.

2.3. Describe SCADA and its functions. Also name the SCADA device that performs data acquisition and switching enablement in the field.

2.4. Write the equation that the least square method aims to minimize. Describe each variable.

2.5. Solve the linear system example in Section 2.3.3.1 given the following covariance matrix and measured values:

$$[R] = \begin{bmatrix} 0.005 & \\ & 0.02 \end{bmatrix}, \quad Z^{\text{meas}} = \begin{bmatrix} M_1 \\ M_2 \end{bmatrix} = \begin{bmatrix} 120 \\ 50 \end{bmatrix}$$

2.6. For the linear example of Section 2.3.3.2, obtain the values of θ_1 and θ_2 given the following equations and values:

$$M_{12} = f_1(\theta_1, \theta_2) = 5\theta_1 - 4\theta_2$$
$$M_{13} = f_2(\theta_1, \theta_2) = 3\theta_1$$
$$M_{32} = f_3(\theta_1, \theta_2) = -6\theta_2$$

$$Z^{\text{meas}} = \begin{bmatrix} Z_{12}^{\text{Meas}} \\ Z_{13}^{\text{Meas}} \\ Z_{23}^{\text{Meas}} \end{bmatrix} = \begin{bmatrix} 0.5 \\ 0.2 \\ 0.4 \end{bmatrix} \quad \text{and} \quad [R] = \begin{bmatrix} 0.001 & & \\ & 0.0001 & \\ & & 0.004 \end{bmatrix}$$

2.7. For the nonlinear system example of Section 2.3.3.3, obtain the V_2 and δ_2 values for the second and third iterations using the same initial conditions.

2.8. For the nonlinear system example of Section 2.3.3.3, what are the V_2 and δ_2 values for the first iteration using the following initial condition?

$$\begin{bmatrix} x_1 \\ x_2 \end{bmatrix} = \begin{bmatrix} \delta_2 \\ V_2 \end{bmatrix} = \begin{bmatrix} 0.5 \\ 1.05 \end{bmatrix}$$

$$\begin{bmatrix} \delta_2 \\ V_2 \end{bmatrix} = \begin{bmatrix} -0.11938 \\ 0.78277 \end{bmatrix}$$

CHAPTER 3

POWER SYSTEM SCENARIO ANALYSIS

3.1 OPERATOR FUNCTION IN POWER SYSTEM SCENARIO ANALYSIS

While monitoring the prevailing operating conditions, an operator is constantly presented with situations in which he needs to make a decision on his next move. Examples of such decisions include responding to an immediate need to come up with a mitigating switching action to resolve a voltage or flow violation or responding to a request to take a piece of equipment out for maintenance within a specified time frame. For the former case, the operator needs to simulate a number of scenarios, each deploying different mitigating actions, and choose the best one among them. In the latter case, the operator needs to simulate the post-outage operating condition and assess its feasibility. Again, he may need to test many scenarios and potentially design a mitigating action if the post-outage scenario ends up being infeasible.

3.2 PROCESS FOR POWER SYSTEM SCENARIO ANALYSIS

As part of the system operating order or process book, utilities document closely what the operational limits and thresholds are for their major substations, transmission lines, and transformers. They also document broadly the type of mitigating actions operators need to take to return the system to within these operating limits is needed to remove a problem such as voltage or power flow violations.

Let us take a high-voltage substation in an urban area where it supplies a substantial load. Operators have written procedures on exact voltage profile threshold that this substation would need to be operated within (e.g., 1.02–1.05 p.u.). During the peak periods, the voltage is depressed, while during the offpeak period, the voltage is increased. Operators need to take mitigating actions at their disposal to bring the voltage to within the range. During peak periods, they may switch some capacitors in or switch some reactors out. On the other hand, during the offpeak periods, they may have to switch some capacitors out or switch some reactors in. They may also have other mitigating actions such as changing the voltage setting on

Practical Power System Operation, First Edition. Ebrahim Vaahedi.
© 2014 The Institute of Electrical and Electronics Engineers, Inc. Published 2014 by John Wiley & Sons, Inc.

a condenser or a generator. Similar situations can happen for transmission lines and transformers in that their actual power flow is limited by their thermal limits and other operational limits which will be discussed later in this book. Again, there are written procedures on what the power flow limits are and what mitigating actions the operator needs to resort to bring the flows to within the acceptable thresholds. Again as an example, let us assume that there is a 500 kV line bringing a remote generation to the load center. The operating book specifies a maximum limit of 900 MW on this line. If the flow on this line exceeds the 1100 MW limit, then there are written procedures in the operating order book on what mitigating actions the operator can resort to. Examples of mitigating actions include changing the generation pattern in the system or reconfiguring the network.

To identify exactly which of the mitigating actions suggested by the operating book should be used and by how much, operators need to examine different scenarios. For example, to remove a voltage violation, the operator needs to examine different scenarios deploying different combinations of capacitors and reactors identified in the operating book.

3.3 TECHNOLOGY FOR POWER SYSTEM CONTROL

The decision support system required for power system control includes the infrastructure to perform switching actions and a tool for scenario analysis.

3.3.1 Infrastructure for Power System Control

The infrastructure used for implementing switching actions is through SCADA, which was described in Chapter 2. To perform the switching action, operators click on switches that they want to open on the screen. Switching action is moved from the man–machine interface to the RTUs, which activate the relays associated with the switches that need to be opened or closed. The decision support also needs a tool called Power Flow, which is described next.

So let us assume that the operator, through some scenario analysis, decides to perform a switching action. He needs to first examine the system mimic diagram providing the prevailing condition, a magnified section of which is shown in Figure 3.1. This diagram that provides the high-voltage layer of the system is in the breaker node model showing the details of the breaker positions. The mimic diagram provides the generation, load, and power flows on the high-voltage system. The operator will need to zoom on a specific substation or transmission equipment where he wants to perform the switching action. Figures 3.2 and 3.3 show the details of a generating station and a transmission line respectively obtained by the operator selecting a specific station or equipment. Operators use these detailed diagrams to implement a switching action by selecting a device such as a circuit breaker or a switch and performing the action on the device. Figure 3.4 shows the transmission diagram of 3.3 when the operators decide to open a device. The operator will be presented with a dialogue box confirming his intention and the actions he can take on this device.

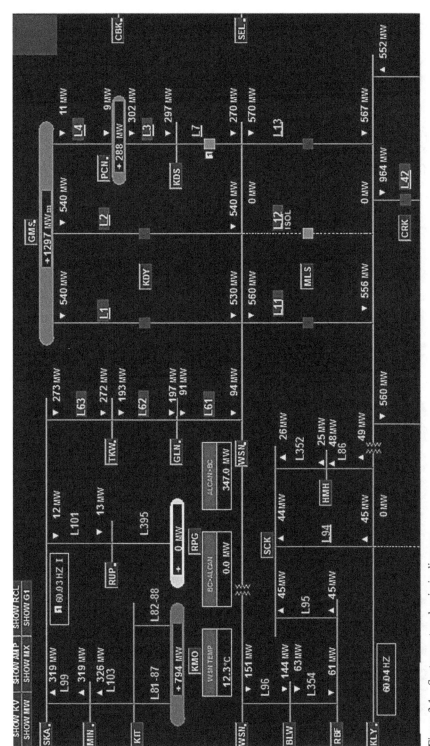

Figure 3.1 System network mimic diagram.

23

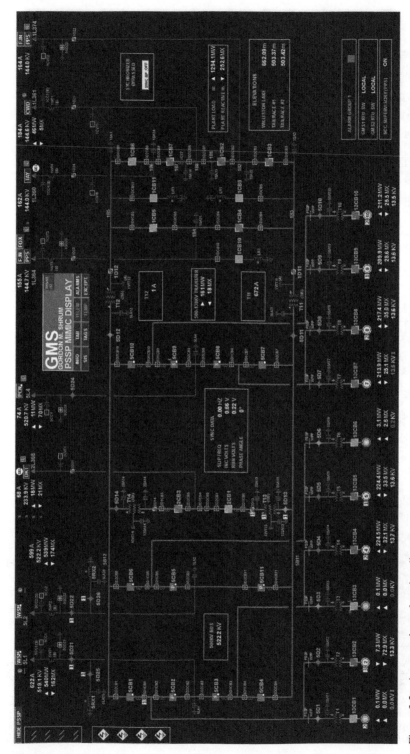

Figure 3.2 A generation substation diagram.

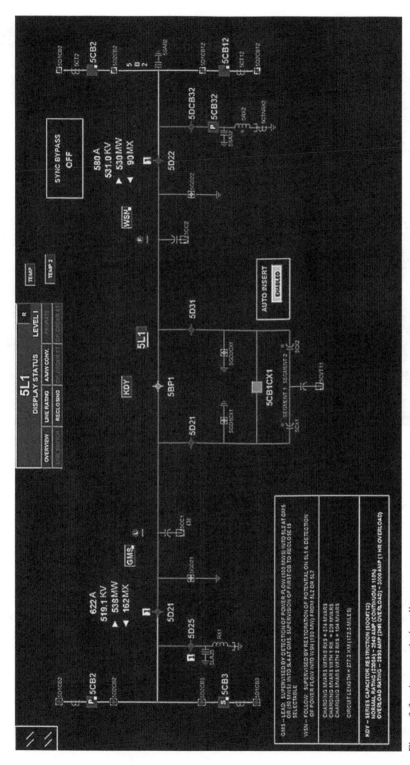

Figure 3.3 A transmission diagram.

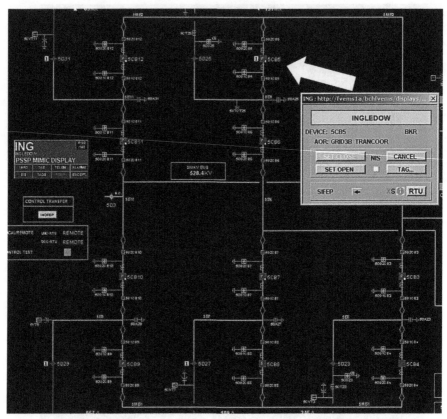

Figure 3.4   Operator attempting to open a circuit breaker with the system responding with a dialogue box confirming his intentions.

Finally, operators are also provided with bus-branch overview diagrams as shown in Figure 3.5. These diagrams do not include the breaker statuses providing a simpler system mimic.

3.3.2   Technology for Power System Scenario Analysis: Power Flow

To establish the system conditions in different scenarios resulting from performing switching actions on the existing system conditions, we need a decision support tool called power flow analysis. The function of a power flow can be described by the following statement:

> Given the load power consumption at all of the buses of the electric power system and the generator power production at each power plant, find the power flow in each line and transformer of the interconnecting network.

The system is assumed to be balanced, allowing a single-phase representation of the system.

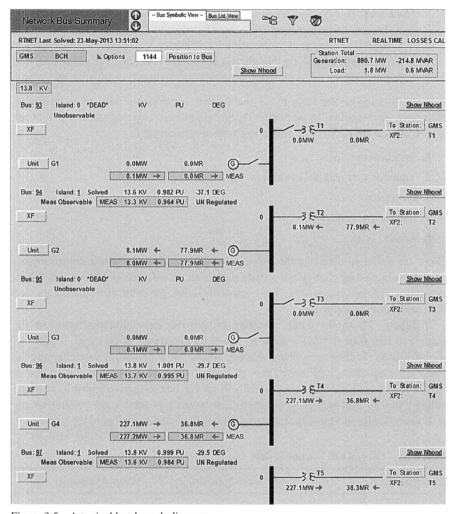

Figure 3.5   A typical bus branch diagram.

3.3.3   System Modeling

Each bus in the system has four quantities: P, Q, V, and Θ. Using the convention that injecting P and Q to the bus is positive and that drawing away P and Q from the system is negative, we need to categorize the system buses according to the availability of the four attributes, P, Q, V, and Θ. As described next, there are three types of buses: PQ, PV, and slack.

PQ Bus: A PQ bus is a load bus with known values for P and Q.

PV Bus: A PV bus is a generator bus with the values of P and V specified. Also, Q for these buses is limited between Q_{max} and Q_{min}.

> ***Slack Bus***: This bus is a reference bus in the system with V and angle specified. It absorbs the system load and generation mismatch. Mathematically, this bus is needed to supply the generation load balance resulting from the losses in the network. In real life though, this bus reflects the action of generation load balance, which takes place by automatic generation control generating units.

In order to analyze the system, all the system components have to be modeled. Normally, as described in Appendix A, the intent is to come up with a π network representation for each of the following system components:

1. Transmission lines
2. Transformers and phase shifters
3. Generators
4. Shunts and condensers
5. Loads

There are also a number of control objectives in a power flow. These control objectives are achieved by changing a control parameter on a device. Table 3.1 lists different control aids in a power flow. The control objective, the controlling parameter, and the control device are given.

Figure 3.6 shows a tap-changing transformer operation that controls the voltage at bus B or the MVar flow to bus B by changing its tap ratio. Figure 3.7 shows a phase shifter or a phase-shifting transformer operation that changes the phase of a tap-changing transformer to control the real power flowing to bus B. Table 3.1 also indicates that a generator's MVar can be changed to adjust its terminal voltage. In a similar way, a synchronous condenser, a generator not generating any real power, is also used to adjust its terminal voltage by changing its MVar generation.

TABLE 3.1   Control aids in a power flow

Control device	Control Parameter	Control Objective
Tap changing transformers	Tap ratio	Voltage, MVar
Phase shifters	Phase	Power flow in a line
Generators	Reactive power	Generator terminal voltage
Synchronous condensers	Reactive power	Voltage
Switchable capacitors, reactors, and SVC	Reactive power	Voltage
Area control	Slack power in areas	Interchange P

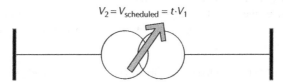

$$V_2 = V_{scheduled} = t \cdot V_1$$

Figure 3.6   Tap changing transformer changing its tap ratio t to set Bus 2 voltage at a scheduled value $V_{scheduled}$.

Figure 3.7   Phase shifting transformer changing its tap ratio angle t to set power flow at a scheduled value $P_{\text{scheduled}}$.

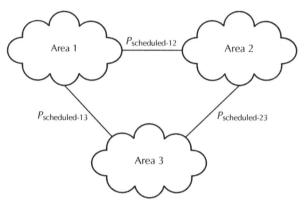

Figure 3.8   Area control ensures that the power flows between different areas are set at scheduled values.

Switchable capacitors, switchable reactors, and static var compensators (SVC) can also be used to control the voltage by changing their generated reactive power. Finally, Figure 3.8 shows the area control operation in which the power generated at the slack buses in different areas are adjusted to control the interchange power flow between the areas.

3.3.4   Power Flow Techniques

The power flow is a set of nonlinear equations that represent the balanced steady-state operation of the power system [9,10].

Two sets of equations, the network equation $I = Y \cdot V$ and the power equation $S = V \cdot I^*$, should be solved together to obtain V and I. Different iterative methods have been developed for solving these equations. These methods differ in characteristics such as the storage requirement, time of solution, robustness, etc. A brief review of different power flow methods have been given later.

For a bus k, the network equation $I = Y \cdot V$ and power equation $S = V \cdot I^*$ can be written as

$$I_k = \frac{(P_k - JQ_k)}{V_k{}^*} \tag{3.1}$$

$$I_k = Y_{kk}V_k + \sum_{\substack{i=1 \\ i \neq k}}^{n} Y_{ki} \cdot V_i \tag{3.2}$$

where I_k, V_k, P_k, and Q_k are injected current, voltage, injected active power, and injected reactive power respectively at bus k. Please note the convention used in these equations is that injected I, P, and Q values are positive, whereas supplied values out of the network are negative. Also, Y_{ki} is the element ki of the Y matrix.

By eliminating I_k from the aforementioned two equations, the following equation is obtained:

$$V_k = \frac{P_k - JQ_k}{Y_{kk}V_k{}^*} - \frac{1}{Y_{kk}} \sum_{\substack{i=1 \\ i \neq k}}^{n} Y_{ki}V_i \tag{3.3}$$

This complex equation can be broken into two real equations to obtain two unknowns of voltage magnitude and phase (load bus).

3.3.4.1 Gauss, Gauss–Seidel These are nonderivative methods using algebraic equations of the form:

$$V^{n+1} = f(V^n) \tag{3.4}$$

where n represents the iteration number. As a simple example, the following function is solved using the Gauss–Seidel method:

$$f(x) = x + \sin(x) - 2 = 0.$$

To solve this equation, it is written in the form of

$$x = 2 - \sin(x)$$

A starting value of $x = 0$ is selected, and new values are obtained:

$$x^1 = 2 - \sin(0) = 2$$

$$x^2 = 2 - \sin(2) = 1.09$$

$$x^3 = 2 - \sin(1.09) = 1.113$$

$$x^4 = 2 - \sin(1.113) = 1.103$$

$$x^{13} = 2 - \sin(1.0606) = 1.10606$$

Similar to the example provided earlier, Equation 3.3 can be written as

$$V_k^{m+1} = \frac{P_k - JQ_k}{(V_k^m)^* Y_{kk}} - \frac{1}{Y_{kk}} \sum_{\substack{i=1 \\ 1 \neq k}}^{n} Y_{ki} V_i^m \tag{3.5}$$

This formulation is called the Gauss formulation. Equation 3.5 can also be written in a slightly different way resulting in the following Gauss–Seidel formulation:

$$V_k^{m+1} = \frac{P_k - JQ_k}{(V_k^m)^* Y_{kk}} - \frac{1}{Y_{kk}} \sum_{i=1}^{k-1} Y_{ki} V_i^{m+1} - \frac{1}{Y_{kk}} \sum_{i=k+1}^{n} Y_{ki} V_i^m \tag{3.6}$$

It should be noted that the Gauss–Seidel formulation of Equation 3.6 is obtained by breaking the second term of the Gauss formulation, $\left(-1/Y_{kk} \sum_{\substack{i=1 \\ 1 \neq k}}^{n} Y_{ki} V_i^m \right)$, into two terms. The first term $\left(-1/Y_{kk} \sum_{i=1}^{k-1} Y_{ki} V_i^{m+1} \right)$ corresponds to the buses for which new voltages have been calculated in iteration $m + 1$. The second term $\left(-1/Y_{kk} \sum_{i=k+1}^{n} Y_{ki} V_i^m \right)$ corresponds to buses for which no new values are available in iteration $m + 1$.

Either of these two equations is solved iteratively until the correction becomes less than a threshold value.

For a generator bus that has a set value for its voltage magnitude, the following equation is used to obtain Q_k:

$$Q_k = -\mathrm{Im}\ g \left[V_k^* \sum_{i=1}^{n} Y_{ki} V_i \right] \tag{3.7}$$

This value of Q_k is used in Equations 3.5 or 3.6 to get the new values for generator voltage magnitude and phase. The voltage magnitude is ignored, and the phase value is used. If the generator Q becomes larger than its Q_{max} or smaller than Q_{min}, then it will be treated like a load bus with its Q set at the limiting value ($Q = Q_{max}$ or $Q = Q_{min}$).

To expedite the solution, an acceleration factor is used in obtaining the new values:

$$V_k^{new} = V_k^{old} + \alpha \left(V_k^{new} - V_k^{old} \right) \tag{3.8}$$

Example 3.1
Solve the following equations using the Gauss and the Gauss–Seidel methods with the initial conditions of $x_1 = 0$ and $x_2 = 0$.

$$x_1 = 2 - \sin(x_1) - x_2$$
$$x_2 = 3 - x_1 - \sin(x_2)$$

Solution for Gauss method:

$$x_1(1) = 2 - \sin(0) - 0 = 2$$
$$x_2(1) = 3 - 0 - \sin(0) = 3$$

$$x_1(2) = 2 - \sin(2) - 3 = -1.9093$$
$$x_2(2) = 3 - 2 - \sin(3) = 0.8589$$

Solution for Gauss–Seidel method:

$$x_1(1) = 2 - \sin(0) - 0 = 2$$
$$x_2(1) = 3 - 2 - \sin(0) = 1$$

$$x_1(2) = 2 - \sin(2) - 1 = 0.09070$$
$$x_2(2) = 3 - 0.09070 - \sin(1) = 2.0678$$

Example 3.2

Solve the following two-bus system power flow using the Gauss method (Figure 3.9).

We first develop the Y_{bus} matrix using Y_{12}:

$$Y_{bus} = \begin{bmatrix} Y_{11} & Y_{12} \\ Y_{21} & Y_{22} \end{bmatrix} = \begin{bmatrix} J50 & -J50 \\ -J50 & J50 \end{bmatrix}$$

Using Equation 3.5, we get

$$V_2 = \frac{P_2 - JQ_2}{Y_{22}(V_2)^*} - \frac{1}{Y_{22}} Y_{21} V_1 {}^*$$

$$V_2{}^{n+1} = \frac{-0.5 - J(-0.5)}{-J50(V_2{}^n)^*} - \frac{1}{-J50} J50(1.06)$$

$$V_2{}^{n+1} = \frac{-0.5 - J(-0.5)}{-J50(V_2{}^n)^*} + (1.06)$$

To solve this equation, we start with $V_2{}^0 = 1 \angle 0.05$

$$V_2{}^1 = \frac{-0.5 - J(-0.5)}{-J50(1 \angle -0.05)} + (1.06) = 1.05 - j0.01$$

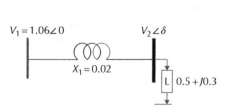

Figure 3.9   The two-bus system.

We continue with this process until the change in V_2 from one iteration to the next becomes less than a threshold.

To expedite the solution, an acceleration factor is used in obtaining the new values.

3.3.4.2 Newton–Raphson

This method is based on the Taylor expansion of a function $f(x+dx)=0$ around the point x:

$$f(x+\Delta x) = f(x) + f'\Delta x + - + - + = 0 \tag{3.9}$$

We want to find a Δx that makes the aforementioned equation equal to zero:

$$\Delta x = -\frac{f(x)}{f'(x)} \tag{3.10}$$

Since $\Delta x = x(\text{new}) - x(\text{old})$,

$$x(\text{new}) = x(\text{old}) - \frac{f(x)}{f'(x)} \tag{3.11}$$

This equation can be described graphically. Figure 3.10 shows a nonlinear function $f(x)$ for which the value of X needs to be found for which $f(X)=0$. Assuming that the starting point is at X_0, then $f(x)$ can be approximated by the tangent line to $f(x)$ at (X_0, Y_0). The new solution X_1 can be found from triangle ABC knowing the slope of the tangent is equal to $f'(X_0)$. Or,

$$f'(X_0) = \frac{[f(X_1) - f(X_0)]}{(X_1 - X_0)} \tag{3.12}$$

Since $f(X_1)=0$, then Equation 3.12 can be expressed as

$$\Delta x = X_1 - X_0 = -\frac{f(X_0)}{f'(X_0)} \tag{3.13}$$

This represents a restatement of Equation 3.10.

In the power flow solution, the aforementioned equation is written in the form

$$X^{n+1} = X^n + \left[J\right]^{-1} \cdot f(X^n) \tag{3.14}$$

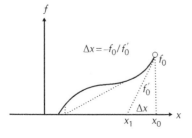

Figure 3.10   Graphical representation of Newton method.

where [J] is the Jacobian matrix in the form of

$$[J] = [f'(x)] = \left[\frac{df}{dx}\right] \tag{3.15}$$

To clarify this method, the simple example given before is solved here using the Newton–Raphson technique:

$$f(x) = x + \sin(x) - 2 = 0$$

$$f'(x) = 1 + \cos(x)$$

$$\Delta x = x^{k+1} - x^k = -\frac{f(x^k)}{f(x^k)'}$$

$$x^0 = 0$$

$$\Delta x = -\frac{(0 + \sin(0) - 2)}{(1 + \cos(0))} = x^1 - x^0$$

$$x^1 = x^0 + 1 = 1$$

$$\Delta x = -\frac{((1 + \sin(1) - 2))}{(1 + \cos(1))} = x^2 - x^1$$

$$x^2 = 1.103$$

$$\Delta x = -\frac{(1.103 + \sin((1.103)) - 2)}{(1 + \cos(1.103))} = x^1 - x^0$$

$$x^3 = 1.10603$$

This example shows that the Newton–Raphson method reaches the same value that Gauss–Seidel obtained in 3 iterations as opposed to 11 iterations by Gauss–Seidel.

To apply this method to power flow solution, Equation 3.3 is restated as

$$P_k - JQ_k = V_k^* \sum_{i=1}^{n} (G_{ki} + JB_{ki}) V_i \tag{3.16}$$

where $G_{ki} + JB_{ki}$ is the ki element of the Y_{bus} matrix.

This equation can be written in the form of

$$P_k - JQ_k = |V_k| \sum_{i=1}^{n} (G_{ki} + JB_{ki}) |V_i| \left[\cos(\vartheta_i - \vartheta_k) + J\sin(\vartheta_i - \vartheta_k)\right] \tag{3.17}$$

From these two equations, values are obtained for P_k and Q_k at each bus:

$$P_k = |V_k| \sum_{i=1}^{n} |V_i| \left[G_{ki} \cos(\vartheta_k - \vartheta_i) + B_{ki} \cdot \sin(\vartheta_k - \vartheta_i) \right] \tag{3.18}$$

$$Q_k = |V_k| \sum_{i=1}^{n} |V_i| \left[G_{ki} \sin(\vartheta_k - \vartheta_i) - B_{ki} \cdot \cos(\vartheta_k - \vartheta_i) \right] \tag{3.19}$$

We write Equations 3.18 and 3.19 in the following forms:

$$f_1 = P_k - |V_k| \sum_{i=1}^{n} |V_i| \left[G_{ki} \cos(\vartheta_k - \vartheta_i) + B_{ki} \cdot \sin(\vartheta_k - \vartheta_i) \right] = 0 \tag{3.20}$$

and

$$f_2 = Q_k - |V_k| \sum_{i=1}^{n} |V_i| \left[G_{ki} \sin(\vartheta_k - \vartheta_i) - B_{ki} \cdot \cos(\vartheta_k - \vartheta_i) \right] = 0 \tag{3.21}$$

We can now solve these two sets of equations belonging to change in P and Q. We call P_k and Q_k, $\mathbf{P}_{sp}$ ($P_{specified}$) and $\mathbf{Q}_{sp}$ ($Q_{specified}$) respectively since these values are specified at the bus and we are seeking values for voltages and angles to make the aforementioned equation equal to zero.

From Equations 3.20 and 3.21, the error for P and Q at each bus is calculated and given in a matrix form as

$$x(\text{new}) = x(\text{old}) - \frac{f(x)}{f(x)'} \tag{3.22}$$

So we need to establish $f(x)'$ the derivative of vector $f(x)$ given in Equation 3.23 with respect to V and ϑ:

$$\begin{bmatrix} f_1 \\ f_2 \end{bmatrix} = \begin{bmatrix} \Delta P \\ \Delta Q \end{bmatrix} = \begin{bmatrix} \partial P / \partial \Theta & \partial P / \partial V \\ \partial Q / \partial V & \partial Q / \partial V \end{bmatrix} \cdot \begin{bmatrix} \Delta \Theta \\ \Delta V \end{bmatrix} \tag{3.23}$$

where ΔP is a vector containing a number of f_1 functions given in Equation 3.20 and ΔQ is a vector containing a number of f_2 functions given in Equation 3.21, respectively. These sets of equations are solved iteratively until the corrections become less than a threshold value.

For a generator bus that has a fixed value for its voltage magnitude, only P needs to be specified. If the generator Q violates its limits becoming larger than its Q_{max} or smaller than Q_{min}, then it will be treated like a load bus with its Q forced at $Q = Q_{max}$ or $Q = Q_{min}$ respectively.

Example 3.3

Solve the two-bus system of using the Newton–Rapson method.

Solution

Using the system parameters, we can write Equations 3.18 and 3.19 as

$$P_2 = |V_2| \sum_{i=1}^{2} |V_i| \big[B_{2i} \cdot \sin(\vartheta_2 - \vartheta_i) \big] = B_{21}V_1V_2 \sin(\vartheta_2 - \vartheta_1) + B_{22}V_2^2 \sin(\vartheta_2 - \vartheta_2)$$

$$Q_2 = |V_2| \sum_{i=1}^{n} |V_i| \big[-B_{2i} \cdot \cos(\vartheta_2 - \vartheta_i) \big] = -B_{21}V_1V_2 \cos(\vartheta_2 - \vartheta_1) - B_{22} V_2^2 \cos(\vartheta_2 - \vartheta_2)$$

This gives

$$P_2 = B_{21}V_1V_2 \sin(\vartheta_2) + B_{22}V_2^2 \sin(\vartheta_2 - \vartheta_2) = B_{21}V_1V_2 \sin(\vartheta_2)$$

$$Q_2 = -B_{21}V_1V_2 \cos(\vartheta_2 - \vartheta_1) - B_{22}V_2^2 \cos(\vartheta_2 - \vartheta_2) = -B_{21}V_1V_2 \cos(\vartheta_2) - B_{22}V_2^2$$

Knowing that $B_{21} = -1/X_{21}$ and $B_{22} = 1/X_2$, we get

$$P_2 = \frac{V_1 V_2}{X_{21}} \sin(\vartheta_2)$$

$$Q_2 = \frac{V_1 V_2}{X_{21}} \cos(\vartheta_2) - \frac{V_2^2}{X_{21}}$$

$$f_1 = P_{2\text{spec}} - \frac{V_1 V_2}{X_{21}} \sin(\vartheta_2) = 0$$

$$f_2 = Q_{2\text{spec}} - \left[\frac{V_1 V_2}{X_{21}} \cos(\vartheta_2) - \frac{V_2^2}{X_{21}} \right] = 0$$

$$\begin{bmatrix} f_1 \\ f_2 \end{bmatrix} = \begin{bmatrix} \Delta P_2 \\ \Delta Q_2 \end{bmatrix} = \begin{bmatrix} \partial P_2 / \partial \Theta_2 & \partial P_2 / \partial V_2 \\ \partial Q_2 / \partial_2 V & \partial Q_2 / \partial V_2 \end{bmatrix} \cdot \begin{bmatrix} \Delta \Theta_2 \\ \Delta V_2 \end{bmatrix}$$

$$= \begin{bmatrix} \dfrac{V_1 V_2}{X_{21}} \cos(\vartheta_2) & \dfrac{V_1}{X_{21}} \sin(\vartheta_2) \\ \dfrac{V_1 V_2}{X_{21}} \sin(\vartheta_2) & -\dfrac{V_1}{X_{21}} \cos(\vartheta_2) + \dfrac{2V_2}{X_{21}} \end{bmatrix} \begin{bmatrix} \Delta \Theta_2 \\ \Delta V_2 \end{bmatrix}$$

We start by assuming initial values for $V_2 = 1$ and $\Theta_2 = 0.05$ and then continue calculation.

We use the aforementioned equation to calculate $\begin{bmatrix} \Delta \Theta_2 \\ \Delta V_2 \end{bmatrix}$

$$\begin{bmatrix} -0.5 - 2.648896 \\ -0.3 - (-2.93376) \end{bmatrix} = \begin{bmatrix} 52.93376 & 2.648896 \\ 2.648896 & 47.06624 \end{bmatrix} \begin{bmatrix} \Delta \Theta_2^1 \\ \Delta V_2^1 \end{bmatrix}$$

$$\begin{bmatrix} \Delta\Theta_2^1 \\ \Delta V_2^1 \end{bmatrix} = \begin{bmatrix} 52.93376 & 2.648896 \\ 2.648896 & 47.06624 \end{bmatrix}^{-1} \begin{bmatrix} -0.5 - 2.648896 \\ -0.3 - (-2.93376) \end{bmatrix} = \begin{bmatrix} -0.06246 \\ -0.05974 \end{bmatrix}$$

$$\begin{bmatrix} \Theta_2^1 \\ V_2^1 \end{bmatrix} = \begin{bmatrix} \Theta_2^0 \\ V_2^0 \end{bmatrix} + \begin{bmatrix} -0.06246 \\ 0.05947 \end{bmatrix} = \begin{bmatrix} -0.01246 \\ 1.05947 \end{bmatrix}$$

We continue with this process until the changes $\begin{bmatrix} \Delta\Theta_2 \\ \Delta V_2 \end{bmatrix}$ become less than a threshold.

3.3.4.3 Decoupled, Fast Decoupled
By simplifying the Jacobian matrix, eliminating P–V and Q–Θ relationships, the decoupled formulation is obtained. The power flow formulation is simplified to

$$\begin{bmatrix} \Delta P \\ \Delta Q \end{bmatrix} = \begin{bmatrix} \partial P / \partial\Theta & 0 \\ 0 & \partial Q / \partial V \end{bmatrix} \cdot \begin{bmatrix} \Delta\Theta \\ \Delta V \end{bmatrix} \tag{3.24}$$

The elements of the Jacobian matrix are voltage-dependent and have to be calculated at each iteration. The advantage of decoupled formulation is that the storage requirement is reduced by a factor of 35–40% and the computation per iteration by a factor of 10–20% [11,12].

Further simplifications in the decoupled formulation were suggested by Stott and Alsac [13] to produce a constant Jacobian matrix. The power flow formulation is reduced to solving a set of linear equations with constant parameters:

$$\Delta P = (VB'V)\Delta\Theta \tag{3.25}$$

$$\Delta Q = (VB''V)\Delta V \tag{3.26}$$

where B' and B'' are equal to $-B$, the network susceptance matrix. The B matrix, which is the imaginary part of the $Y = G + JB$, only contains real values and can be formed similar to the Y matrix. The following simplifications further improve the convergence rate of the iterative process [13].

1. The network elements affecting the reactive power flows such as shunt reactances are deleted from B'. Similarly, phase shifters are deleted from B''.
2. Equations 3.25 and 3.26 are converted to the following equations:

$$\frac{\Delta P}{V} = B'\Delta\Theta \tag{3.27}$$

$$\frac{\Delta Q}{V} = B''\Delta V \tag{3.28}$$

This method is fast and converges reliably. Poor convergence can result if the assumption of high X/R ratios is false. Haley and Ayres [14] suggested a method to resolve the problem of poor convergence of the fast decoupled power flow by an elegant transformation of voltages and power injections. The new algorithm, which is called the super-decoupled power flow, can efficiently handle the systems with low X/R ratios.

The method given in Reference [15], which was obtained by a mistake in developing a fast decoupled method, is different from the original method in that instead of removing the impact of series resistances in B', it deletes them in B''.

Example 3.4

Solve Example 3.1 using decoupled and fast decoupled Newton–Raphson.

Solution

Using the Jacobian matrix in and eliminating the P–V and Q–Θ relationships, we obtain the following Jacobian matrix for the decoupled Newton–Raphson:

$$
\begin{bmatrix} \partial P_2 / \partial \Theta_2 & 0 \\ 0 & \partial Q_2 / \partial V_2 \end{bmatrix} = \begin{bmatrix} \dfrac{V_1 V_2}{X_{21}} \cos(\vartheta_2) & 0 \\ 0 & -\dfrac{V_1}{X_{21}} \cos(\vartheta_2) + \dfrac{2V_2}{X_{21}} \end{bmatrix}
$$

resulting in the following relationship:

$$
\begin{bmatrix} f_1 \\ f_2 \end{bmatrix} = \begin{bmatrix} \Delta P_2 \\ \Delta Q_2 \end{bmatrix} = \begin{bmatrix} \partial P_2 / \partial \Theta_2 & 0 \\ 0 & \partial Q_2 / \partial V_2 \end{bmatrix} \cdot \begin{bmatrix} \Delta \Theta_2 \\ \Delta V_2 \end{bmatrix}
$$

$$
= \begin{bmatrix} \dfrac{V_1 V_2}{X_{21}} \cos(\vartheta_2) & 0 \\ 0 & -\dfrac{V_1}{X_{21}} \cos(\vartheta_2) + \dfrac{2V_2}{X_{21}} \end{bmatrix} \begin{bmatrix} \Delta \Theta_2 \\ \Delta V_2 \end{bmatrix}
$$

We start by assuming initial values for $V_2 = 1$ and $\Theta_2 = 0.05$ and then continue calculation.

We use the aforementioned equation to calculate $\begin{bmatrix} \Delta \Theta_2 \\ \Delta V_2 \end{bmatrix}$

$$
\begin{bmatrix} -0.5 - 2.648896 \\ -0.3 - (-2.93376) \end{bmatrix} = \begin{bmatrix} 52.93376 & 0 \\ 0 & 47.06624 \end{bmatrix} \begin{bmatrix} \Delta \Theta_2^1 \\ \Delta V_2^1 \end{bmatrix}
$$

$$
\begin{bmatrix} \Delta \Theta_2^1 \\ \Delta V_2^1 \end{bmatrix} = \begin{bmatrix} 52.93376 & 0 \\ 0 & 47.06624 \end{bmatrix}^{-1} \begin{bmatrix} -0.5 - 2.648896 \\ -0.3 - (-2.93376) \end{bmatrix} = \begin{bmatrix} -0.05949 \\ 0.055959 \end{bmatrix}
$$

$$
\begin{bmatrix} \Theta_2^1 \\ V_2^1 \end{bmatrix} = \begin{bmatrix} \Theta_2^0 \\ V_2^0 \end{bmatrix} + \begin{bmatrix} -0.05949 \\ 0.055959 \end{bmatrix} = \begin{bmatrix} -0.00949 \\ 1.055959 \end{bmatrix}
$$

We continue with this process until the changes $\begin{bmatrix} \Delta \Theta_2 \\ \Delta V_2 \end{bmatrix}$ become less than a threshold.

Solution

From Equations 3.27 and 3.28, we obtain

$$\Delta\Theta = [B']^{-1}\left(\frac{\Delta P}{V}\right)$$

$$\Delta V = [B'']^{-1}\left(\frac{\Delta Q}{V}\right)$$

Knowing that

$$Y_{bus} = G + JB = \begin{bmatrix} Y_{11} & Y_{12} \\ Y_{21} & Y_{22} \end{bmatrix} = \begin{bmatrix} J50 & -J50 \\ -J50 & J50 \end{bmatrix}$$

The B matrix for this example becomes

$$B = \begin{bmatrix} 50 & -50 \\ -50 & 50 \end{bmatrix}$$

If we now ignore the terms associated with Bus 1 because it is the slack bus, then we get

$$B = 50$$

Assuming initial values for $V_2 = 1$ and $\Theta_2 = 0.05$, we obtain the following values:

$$\Delta\Theta_2^1 = [B']^{-1}\left(\frac{\Delta P_2}{V_2}\right) = [-50]^{-1}\left(\frac{-0.5 - 2.648896}{1.0}\right) = -0.06298$$

$$V_2^1 = [B'']^{-1}\left(\frac{\Delta Q_2}{V_2}\right) = [-50]^{-1}\left(\frac{-0.3 - (-2.93376)}{1.0}\right) = 0.05276$$

$$\begin{bmatrix} \Theta_2^1 \\ V_2^1 \end{bmatrix} = \begin{bmatrix} \Theta_2^0 \\ V_2^0 \end{bmatrix} + \begin{bmatrix} -0.06298 \\ 0.05276 \end{bmatrix} = \begin{bmatrix} -0.01298 \\ 1.052675 \end{bmatrix}$$

Again, we continue with this process until the changes $\begin{bmatrix} \Delta\Theta_2 \\ \Delta V_2 \end{bmatrix}$ become less than a threshold.

3.3.4.4 DC Power Flow These methods use an approximate power flow model and are employed when only an approximate solution is required.

DC power flow formulation is obtained from Equation 3.18 when the line resistances are ignored ($G_{Ki} = 0$) and voltages assumed to be unity ($V_i = 1$), resulting in a formulation between P and θ in the form of

$$P_k = \sum_{i=1}^{n}\left[B_{ki} \cdot \sin(\vartheta_k - \vartheta_i)\right] \tag{3.29}$$

where P_k is the injected power at bus k.

Assuming that $(\vartheta_k - \vartheta)$ is small, we can approximate $\sin(\vartheta_k - \vartheta_i)$ as

$$\sin\left(\vartheta_k - \vartheta_i\right) = \left(\vartheta_k - \vartheta_i\right) \tag{3.30}$$

Substituting for $\sin(\vartheta_k - \vartheta_i)$ from Equation 3.30 into Equation 3.29, we get

$$P_k = \sum_{i=1}^{n}\left[B_{ki} \cdot \left(\vartheta_k - \vartheta_i\right)\right] = \sum_{i=1}^{n}\left[-B_{ki} \cdot \left(\vartheta_i\right)\right]^{\text{1}}$$

which can be turned into the following form:

$$P = -[B][\vartheta] \tag{3.31}$$

where P and ϑ are vectors of power injections and angles at different buses and B is the susceptance matrix described earlier.

Example 3.5
Solve Example 3.1 using a DC power flow.

Solution
The Y matrix for Example 3.1 is

$$Y_{\text{bus}} = G + JB = \begin{bmatrix} Y_{11} & Y_{12} \\ Y_{21} & Y_{22} \end{bmatrix} = \begin{bmatrix} J50 & -J50 \\ -J50 & J50 \end{bmatrix}$$

So the B matrix for this example is

$$B = \begin{bmatrix} 50 & -50 \\ -50 & 50 \end{bmatrix}$$

If we now ignore the terms associated with Bus 1 because it is the slack bus, then we get

$$B = 50$$
$$P_2 = -B\theta_2 = -50\theta_2$$
$$-0.5 = -B\theta_2 = -50\theta_2$$
$$\theta_2 = 0.01$$

3.3.4.5   *Comparison of Power Flow Methods*   From the power flow methods described earlier, the following methods are the ones currently used:

1. Newton–Raphson
2. Gauss–Siedel
3. Fast decoupled
4. Modified version of fast decoupled [14]
5. DC

---

[1] We are not assuming $\vartheta_k = 0$. The $B_{ki} \cdot (\vartheta_k)$ terms on the left side of the equation is compensated by the right-side equation when $i = k$.

The Gauss–Siedel method is a reliable method for starting a power flow solution as it is very tolerant of poor initial voltage and reactive power conditions. The method has a slow convergence and can become nonconvergent for some stress cases. This method has low computer memory requirement while its computation time increases rapidly with system size. It has also been reported [16] that this method is not tolerant of slack bus location and has difficulty solving in the following situations:

1. Systems with more than 70° angles.
2. Systems containing negative reactance, for example, series compensation.
3. Systems with long and short lines terminating at the same bus with ratios of over 1000.
4. Systems of long radial shape.
5. Finally, the number of iterations in this method is a function of the accelerating factor.

The Newton–Raphson method is the most reliable power flow method available. It has a very good (quadratic) convergence rate, and its computation time increases linearly with the system size. This method is not tolerant of bad initial conditions and can have problems at the beginning of a power flow solution. This method is suitable for large applications requiring a high degree of accuracy.

The convergence properties of the Gauss–Siedel and Newton methods can be combined in that a solution can be first started by a Gauss–Siedel method, and then after a few iterations, the solution can be switched to the Newton method.

The fast decoupled power flow is much faster than the full Newton–Raphson as it does not have to recalculate the Jacbian matrix. The computer memory requirements are also less with linear convergence rate. These methods are less sensitive than the full Newton–Raphson methods to the initial conditions. The original fast decoupled methods have problems with large R/X ratio lines. The modified versions of the fast decoupled [15,14] are better suited for these methods.

The DC method addresses the active power calculation in a power flow, and it is used in applications requiring an approximate or simplified power flow representation, for example, economic dispatch.

Table 3.2 compares the attributes of different power flow methods.

TABLE 3.2   Comparison of different power flow methods

Method	Starting Performance	Robust	Convergence Rate	Accuracy	CPU
Gauss Siedel	Reliable	Ok	Slow	Good	High
Newton Raphson	So so	Very Good	Good	Good	Linear
Fast decoupled Newton– Raphson	So so	Good	Good	Good	Low
DC	Reliable	Excellent	Good	So so	Low

3.3.5   Factorization

The Newton method requires the inversion of the Jacobian matrix in solving the power flow equations. Factorization is commonly used for solving these equations. Assuming a linear matrix equation in the form

$$A \cdot x = b \tag{3.32}$$

it can be shown that A can be factorized to

$$A = LDU \tag{3.33}$$

where L is the lower triangle matrix, D is a diagonal matrix, and U is the upper triangular matrix. Equation 3.33 can be restated as

$$(LD)U \cdot x = b \tag{3.34}$$

First, the two matrices of LD and U need to be established from A using Gaussian Elimination [17], where LD is a lower triangular matrix. It should be noted that in some formulations L and D are combined and shown as L. Values for $Z = U \cdot x$ can be obtained using forward substitution, meaning that the components of Z are obtained in the sequence of $z_1, z_2, \ldots, z_n$.

Once Z is obtained, the equation $U \cdot x = Z$ is solved using a backward substitution as $U \cdot x$ is an upper triangular matrix, meaning that the components of x are obtained in the order of $x_n, \ldots, x_2, x_1$.

Example 3.6
Solve the following $Ax = b$ equation:

$$\begin{bmatrix} 1 & 2 \\ 3 & 4 \end{bmatrix} \begin{bmatrix} x_1 \\ x_2 \end{bmatrix} = \begin{bmatrix} 1 \\ 2 \end{bmatrix}$$

Solution
We can write the system in form of

$$LUx = b$$

First, A is decomposed into LU using Gaussian elimination. L is a lower matrix defined by LD in Equation 3.34, and U is an upper matrix. We need to eliminate the 3 in the second row by using the top row. So, we perform the following manipulation:

$$\text{Row}_2 = -3\,\text{Row}_1 + \text{Row}_2$$

$$\begin{bmatrix} 1 & 2 \\ 0 & -2 \end{bmatrix} = U$$

We store the negative of the factor, which is $-(-3)$, into L:

$$\begin{bmatrix} 1 & 0 \\ 3 & 1 \end{bmatrix} = L$$

So now the system of equations is changed to the form

$$\begin{bmatrix} 1 & 0 \\ 3 & 1 \end{bmatrix} \underbrace{\begin{bmatrix} 1 & 2 \\ 0 & -2 \end{bmatrix} \begin{bmatrix} x_1 \\ x_2 \end{bmatrix}}_{z} = \begin{bmatrix} 1 \\ 2 \end{bmatrix}$$

We now define z as

$$z = \begin{bmatrix} 1 & 2 \\ 0 & -2 \end{bmatrix} \begin{bmatrix} x_1 \\ x_2 \end{bmatrix}$$

resulting in

$$\begin{bmatrix} 1 & 0 \\ 3 & 1 \end{bmatrix} \begin{bmatrix} z_1 \\ z_2 \end{bmatrix} = \begin{bmatrix} 1 \\ 2 \end{bmatrix}$$

Now we can find z with forward substitution:

$$z_1 = 1$$

$$3z_1 + 1z_2 = 2 \rightarrow z_2 = -1$$

This implies that

$$\begin{bmatrix} 1 & 2 \\ 0 & -2 \end{bmatrix} \begin{bmatrix} x_1 \\ x_2 \end{bmatrix} = \begin{bmatrix} 1 \\ -1 \end{bmatrix}$$

We can finally find x with backward substitution:

$$x_2 = \frac{1}{2}$$

$$x_1 + 2x_1 = 1 \rightarrow x_1 = 0$$

We test these results with the initial equation:

$$\begin{bmatrix} 1 & 2 \\ 3 & 4 \end{bmatrix} \begin{bmatrix} 0 \\ 1/2 \end{bmatrix} = \begin{bmatrix} 1 \\ 2 \end{bmatrix}$$

Example 3.7
Find the inverse of the matrix A:

$$A = \begin{bmatrix} 1 & 2 \\ 3 & 4 \end{bmatrix}$$

Solution
First, we find LD and U matrices using Gaussian elimination [17]. We know that

$$AA^{-1} = I = \begin{bmatrix} 1 & 0 \\ 0 & 1 \end{bmatrix}$$

$$\underbrace{\begin{bmatrix} 1 & 0 \\ 3 & 1 \end{bmatrix}\begin{bmatrix} 1 & 2 \\ 0 & -2 \end{bmatrix}}_{A} A^{-1} = \begin{bmatrix} 1 & 0 \\ 0 & 1 \end{bmatrix} = [b_a, b_b]$$

We form

$$A^{-1} = \begin{bmatrix} a_1 & a_3 \\ a_2 & a_4 \end{bmatrix}$$

So we need to find the solutions to the following equations:

$$\begin{bmatrix} 1 & 2 \\ 3 & 4 \end{bmatrix}\begin{bmatrix} b_{1,3} \\ b_{2,4} \end{bmatrix} = b_i$$

where b_i is a unity set:

$$b_i = \left\{ \begin{bmatrix} 1 \\ 0 \end{bmatrix}, \begin{bmatrix} 0 \\ 1 \end{bmatrix} \right\} = \{b_a, b_b\}$$

Find x_1 and x_2 by solving this system of equations:

$$\begin{bmatrix} 1 & 2 \\ 3 & 4 \end{bmatrix}\begin{bmatrix} a_1 \\ a_2 \end{bmatrix} = \begin{bmatrix} 1 & 0 \\ 3 & 1 \end{bmatrix}\underbrace{\begin{bmatrix} 1 & 2 \\ 0 & -2 \end{bmatrix}\begin{bmatrix} a_1 \\ a_2 \end{bmatrix}}_{z} = \begin{bmatrix} 1 \\ 0 \end{bmatrix} = b_a$$

We perform forward substitution:

$$\begin{bmatrix} 1 & 0 \\ 3 & 1 \end{bmatrix}\begin{bmatrix} z_1 \\ z_2 \end{bmatrix} = \begin{bmatrix} 1 \\ 0 \end{bmatrix}$$

$$z_1 = 1$$

$$3z_1 + 1z_2 = 0 \rightarrow z_2 = -3$$

We then perform backward substitution:

$$\begin{bmatrix} 1 & 2 \\ 0 & -2 \end{bmatrix}\begin{bmatrix} a_1 \\ a_2 \end{bmatrix} = \begin{bmatrix} 1 \\ -3 \end{bmatrix}$$

We can then find the values of a:

$$a_2 = \frac{3}{2}$$

$$a_1 + 2a_1 = 1 \rightarrow x_1 = -3$$

We now find a_3 and a_4:

$$\begin{bmatrix} 1 & 0 \\ 3 & 1 \end{bmatrix} \underbrace{\begin{bmatrix} 1 & 2 \\ 0 & -2 \end{bmatrix} \begin{bmatrix} a_3 \\ a_4 \end{bmatrix}}_{z} = \begin{bmatrix} 0 \\ 1 \end{bmatrix} = b_b$$

Defining z and performing forward substitution, we get

$$\begin{bmatrix} 1 & 0 \\ 3 & 1 \end{bmatrix} \begin{bmatrix} z_3 \\ z_4 \end{bmatrix} = \begin{bmatrix} 0 \\ 1 \end{bmatrix}$$

$$z_3 = 0$$

$$3z_3 + 1z_4 = 0 \rightarrow z_3 = 1$$

Now we perform backward substitution:

$$\begin{bmatrix} 1 & 2 \\ 0 & -2 \end{bmatrix} \begin{bmatrix} a_3 \\ a_4 \end{bmatrix} = \begin{bmatrix} 0 \\ 1 \end{bmatrix}$$

We can finally find x with backward substitution:

$$a_4 = -\frac{1}{2}$$

$$a_3 + 2a_4 = 1 \rightarrow a_3 = 1$$

Finally, the inverse can be found as

$$A^{-1} = \begin{bmatrix} a_1 & a_3 \\ a_2 & a_4 \end{bmatrix} = \begin{bmatrix} -2 & 1 \\ 3/2 & -1/2 \end{bmatrix}$$

Testing the results in the initial equation gives

$$\begin{bmatrix} x_1 \\ x_2 \end{bmatrix} = A^{-1} \begin{bmatrix} 1 \\ 2 \end{bmatrix} = \begin{bmatrix} -2 & 1 \\ 3/2 & -1/2 \end{bmatrix} \begin{bmatrix} 1 \\ 2 \end{bmatrix} = \begin{bmatrix} (-2 \times 1) + (1 \times 2) \\ \left(\frac{3}{2} \times 1\right) + \left(-\frac{1}{2} \times 2\right) \end{bmatrix} = \begin{bmatrix} 0 \\ 1/2 \end{bmatrix} = \begin{bmatrix} x_1 \\ x_2 \end{bmatrix}$$

3.3.6 Sparsity

In factorizing the Jacobian matrix A, which is a sparse matrix (up to 97%), it is important to end up with sparse matrices [13] minimizing the amount of computation required. Sparsity techniques using optimal ordering are essential to this approach for solution of large network equations.

3.3.7 Different Power Flow Scenarios and Applications

Different assumptions and specifications regarding the controls in a power flow can be used to represent different scenarios of the system. These scenarios include prefault and postfault system conditions.

There are basically two types of applications that a power flow solution can be used for: (i) steady-state analysis and (ii) quasi-dynamic studies.

3.3.7.1 Steady-State Applications
The steady-state applications are those that power flows are originally designed for, that is, to calculate the operating conditions given the load power consumption at all of the buses of the electric power system and the generator power production at each power plant and the network. Examples of these applications are

1. Prefault steady-state system conditions to obtain flows and voltages, etc.
2. Postfault steady-state conditions to obtain flows and voltages following a contingency. Controls to be used depend on the steady-state time frame of interest.

In this category, the following applications can be identified:

1. Static security assessment identifying any voltage and flow violations in the prefault and postfault conditions.
2. Voltage stability studies identifying the maximum system margin or load increase.
3. Other system parameter calculations such as losses, generator reactive reserves, etc.

In these applications, the loads are normally represented by constants P and Q as ULTC have sufficient time to operate and retain the load bus voltage. The ULTCs, phase shifters, and area interchange controls also are normally included in the calculations.

3.3.7.2 Quasi-Dynamic Applications
In these applications, power flow solutions are used to portray snapshots of system operating conditions following a contingency.

For example, following a contingency such as loss of generation or load, the system operating conditions can be assessed immediately using an inertial power flow [18] using a power flow solution. Basically in an inertial power flow, the generation or load loss (increase) is distributed among different system generators proportional to their inertia. The argument is that immediately after a contingency the generators in the system will pick up the deficit proportional to their inertia. Similarly, the response of the system after all the generator governors have responded can be established by distributing the loads based on the governor participation on different machines in the system [17]. Finally, the conditions following the operation of the automatic generation control system can be assessed by distributing the generation balance among the regulating units.

QUESTIONS AND PROBLEMS

3.1. List three categories of power buses used in power flow calculations.

3.2. List six power flow control devices with the control parameter and control objective.

3.3. Compare the attributes of different power flow methods.

3.4. What is the significance of Y matrix?

3.5. Describe the difference between decoupled and fast decoupled power flow methods.

3.6. What are the functions of area control, phase-shifting transformer, and tap-changing transformer?

3.7. Name two broad categories of power flow applications.

3.8. For the system shown in Figure 3.11, find the V_2 and δ_2 using Gauss method with two iterations with the following initial conditions and system parameters:

$$V_1 = 1.07\angle 0$$

$$X_{12} = j0.01$$

$$Y_{12} = -j100$$

$$S_1 = 0.9 + j0.2$$

3.9. Solve the following equation using LDU decomposition method:

$$\begin{bmatrix} 1 & 3 \\ 4 & 2 \end{bmatrix}\begin{bmatrix} x_1 \\ x_2 \end{bmatrix} = \begin{bmatrix} 4 \\ 2 \end{bmatrix}$$

3.10. Invert matrix A given in the following using LDU decomposition.

$$A = \begin{bmatrix} 1 & 3 \\ 4 & 2 \end{bmatrix}$$

3.11. Solve the power flow problem of Question 3.8 using the Newton–Raphson method for two iterations:

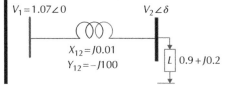

$V_1 = 1.07\angle 0$ $V_2\angle\delta$

$X_{12} = j0.01$
$Y_{12} = -j100$

L $0.9 + j0.2$

Figure 3.11   Two-bus system.

CHAPTER 4

POWER SYSTEM POSTURING: STATIC SECURITY

4.1 OPERATOR'S QUESTION ON POWER SYSTEM POSTURING: STATIC SECURITY

NERC defines the security of a power system by

1. The presence of acceptable operating conditions before and after a contingency
2. The ability of the system to ride through the contingency and to reach the postcontingency operating condition without becoming unstable

The first item characterized by the presence of acceptable operating conditions before and after a contingency is called static security. As per NERC's $n-1$ criteria, the operator needs to ensure that the system condition remains operationally feasible following any credible contingency. The main question that the operator needs to answer is whether the system will be OK after any credible contingency and, to the extent that it is not, what actions he or she needs to take to posture the system so that it becomes operationally feasible.

4.2 PROCESS FOR POWER SYSTEM POSTURING: STATIC SECURITY

The process of posturing the system after any credible condition involves the examination of the system after all credible contingencies establishing whether the system is operating within acceptable thresholds of

- Substation voltage limits
- Transmission line thermal limits
- Generator operating limits

If the system conditions are unacceptable after any contingency, the operator needs to develop a mitigation plan to deal with that contingency. He or she needs to posture the system before the contingency happens or posture the system after the contingency has happened so that the condition become acceptable.

Practical Power System Operation, First Edition. Ebrahim Vaahedi.
© 2014 The Institute of Electrical and Electronics Engineers, Inc. Published 2014 by John Wiley & Sons, Inc.

Again going back to our metaphor of operating a car, a good driver while monitoring the prevailing conditions and ensuring that the system is operating within acceptable thresholds should also proactively posture the car's condition so that it can survive any single event. So to ensure that it can survive an accident, the operator will need to posture the system by reducing its speed before the contingency happens or be able to posture the system after the event by maneuvering the car in a very short period of time.

Through experience and knowledge about their systems, system operators develop remedial action schemes (RAS) for contingencies that normally cause operational infeasibility. These RAS normally involve automatic initiation of system or network changes that need to take place after sensing a contingency. For example, an RAS could involve switching in a number of capacitor banks and switching out a number of reactor banks. These remedial action schemes should be in place to provide comfort to the operator to ensure that the system is OK after any credible contingencies.

4.3 TECHNOLOGY FOR POWER SYSTEM POSTURING: STATIC SECURITY

There are two sets of technologies needed for static security posturing:

1. Contingency analysis application
2. Implementation of remedial action schemes

4.3.1 Contingency Analysis

Security assessment has two functions. The first is the violation detection in the precontingency operating state, which is conducted as part of the system monitoring. The second, much more demanding, function of the security assessment is the detection and evaluation of limit violations after simulating design contingencies. This latter assessment is called contingency analysis.

Contingency analysis is performed on a set of specified design contingency cases. Theoretically, for each contingency, the complete system state should be evaluated to identify and rank those that create violations. The operational planners can then respond to each insecure contingency case and decide to alter the precontingency system operating state to eliminate the violations resulting from the contingency. Alternatively, a control strategy may be developed to alleviate the violations should the contingency occur.

A straightforward way of contingency analysis is to simulate each contingency on the base case to check for operating limit violations. In practice, however, the processing of power flow simulation for a large number of contingency cases requires an enormous computational effort. Any efforts to reduce this computational intensive analysis will result in significant savings in overall computer resources.

Contingency analysis can be divided into three distinct stages:

Contingency definition is the process of enumerating in detail all the contingencies that have a reasonable probability of occurrence or the ones dictated by the security criteria observed. They are specified at the system element level where faults on the system occur. This list normally remains constant; however, the implications of each contingency may vary with system topology and generation. This process is the least time-consuming stage, and techniques to update only those contingencies affected by changes need to be processed.

Contingency selection is the process of shortening the original long list of contingencies by eliminating a vast majority of cases having no violations. The selection is performed to "short-list" the contingencies that need to be studied and thus save time. The selection is performed by either using a fast automated method or by human judgment based on years of operating experience. After the selection, the contingencies are ranked in order of their severity. This process, when automated, is the most time-consuming stage and offers the greatest potential for computational saving.

Contingency evaluation is the process of evaluating the selected contingencies. Evaluation is normally performed using a load flow. With contingency definition and selection preceding this process, the computing requirements for contingency evaluation remains relatively constant and is directly proportional to the number of contingencies evaluated.

Contingency selection and ranking is the most computationally demanding portion of contingency analysis and has been the focus of a lot of developmental effort. Approximate formulations and solutions have to be used in return for speed. From a users' perspective, the combined capability of defining and selecting contingencies in an automated manner greatly enhances their ability to assess many more future conditions.

The power system limits of most interest in contingency analysis are those on branch flows and bus voltages. The accuracy of predicting these limit violations and the turnaround necessary to assess a study condition necessitates compromises based largely on engineering judgment and sheer expediency.

4.3.2 Contingency Definition

The contingencies to be considered for postfault security assessment depend on the utility and their observed reliability criteria. For example, under the North East Power Coordinating Council's criteria, the following contingency types are considered:

1. Double line fault (DLF)
2. Breaker fault (BF)/LG breaker failure
3. Three ph bus fault
4. Three ph line/transformer/generator fault
5. Line end open (LEO)
6. Inadvertent breaker opening (IBO)

A simple method for contingency definition is the manual definition of the system elements to be lost for each contingency by the user. The contingency analysis software will then use this definition to simulate the effect of the contingency on the system.

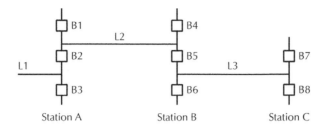

Station "A": Open breakers B1, B2
Station "B": Open breakers B4, B6
Station "C": Open breakers B7, B8

Figure 4.1   Contingency definition for contingency B5 – BF (BF = breaker failure).

For the contingency simulation to be accurate, however, the original definition of each contingency should be updated to reflect the state of the power system at the time of study. To do this, the user will be required to update this definition for every contingency simulation. This requires a high degree of user interface because of the dependency of each contingency on network topology. Furthermore, the time required to set up these conditions for a large number of contingencies to be studied in a specified time horizon makes this method unmanageable.

A more intelligent method of contingency definition requires the users to specify the switching operations that take place for each contingency. The user, with the assumption that all the breakers operate normally, will provide a list of the breakers to open in order to isolate every contingency in the list. A sample contingency definition of this type is shown in Figure 4.1.

Using a topology analyzer software, the breaker operation for each contingency can be translated into element outages automating the manual effort of defining the system element outages. It also determines the additional element losses due to the status of switches and circuit breakers in the precontingency state. These additional element losses are called sympathetic outages. Figure 4.2 shows a typical display available in the contingency analysis application in an energy management for defining contingencies using the circuit breaker statuses.

4.3.3   Contingency Selection

Contingency selection involves screening and ranking of the contingency cases. Screening involves the fast approximate power flow simulation of each contingency case. By monitoring the appropriate postcontingency qualities (flow and/or voltages), the case's severity can be quantified directly for ranking purposes.

4.3.3.1   Contingency Screening   Contingency screening methods are designed to select the contingencies that will cause flow or voltage violation after the contingency. To expedite the process of contingency screening and ranking,

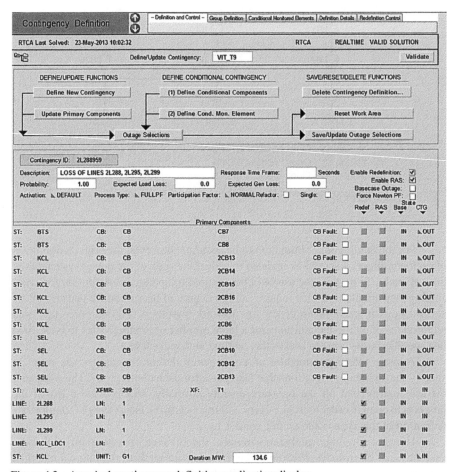

Figure 4.2   A typical contingency definition application display.

approximate calculation methods are used to simulate the operating condition after the contingency. These methods have been summarized next:

- *DC power flow*: This class of methods are only applicable to screening and ranking of flow violations and have no intelligence on voltage violations. The DC power flow approximation in the form of $P = [Bx] \cdot \delta$ is used to calculate the flow conditions after each contingency. A small variation of the aforementioned method is to derive distribution factors from the inverse matrix elements [19]. Following a contingency, the flow change in each monitored branch is given by its distribution factor multiplied by the precontingency flow in the outaged branch. The main benefit of deriving distribution factors is avoiding the calculation of the matrix inversion for every contingency. However, the problem with this approach is that the distribution factors need to be calculated ahead of time using offline calculations and stored for use in real time with huge storage requirements for large-scale systems. For example, let us assume that we have a

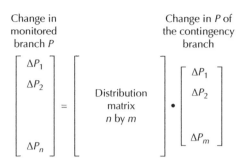

Change in monitored branch P

Change in P of the contingency branch

$$\begin{bmatrix} \Delta P_1 \\ \Delta P_2 \\ \\ \\ \Delta P_n \end{bmatrix} = \begin{bmatrix} \text{Distribution matrix} \\ n \text{ by } m \end{bmatrix} \cdot \begin{bmatrix} \Delta P_1 \\ \Delta P_2 \\ \\ \Delta P_m \end{bmatrix}$$

Figure 4.3   Distribution factor matrix for a case monitoring n branches and m contingencies.

1000 bus system that has about 3000 transmission lines. Let us also assume that we would like to monitor the impact of 2000 outages on all 3000 transmission branches. For this case, we need to calculate a distribution factor matrix of 3000 by 2000 as shown in Figure 4.3. Another drawback with this method is its inability to simulate more complicated contingencies such as bus-split faults.

- **One iteration of fast decoupled power flow**: The fast decoupled power flow method involves the solution of two equations of $\Delta P = [B']\cdot\Delta\delta$ and $\Delta Q = [B'']\cdot\Delta V$ (Eqs. 3.23 and 3.24) consisting of constant matrices B' and B''. One iteration of the first equation $\Delta P = [B']\cdot\Delta\delta$ can be used for expediting the process of contingency screening for active power violations. This method seems to have a better accuracy than the DC power flow method [20]. For voltage violation contingency analysis, however, an iteration of a fast decoupled load flow involving the solution of both equations should be conducted. This method has proven to be unreliable because of the problem nonlinearity and failure to represent voltage control features [20].

  A straightforward method is to build a matrix for each contingency case and invert the matrix for power flow solution. For example, in the example of the DC power flow with $\Delta P = [B']\cdot\Delta\delta$, the $[B']$ matrix needs to be developed for each contingency and inverted. To avoid the inversion for each contingency, the compensation methods or inverse matrix manipulation lemma (IMML) [21] can be used. This method deploys the inverted matrix for the base case and superimposes changes required as a result of the contingency. This removes the need for conducting a matrix inversion for every contingency. Instead, using a limited amount of matrix manipulation, it derives the inverted matrix after a contingency from the inverted matrix for the normal case as given in the equation

$$\left[J_{\text{post}} \right]^{-1} = \left[J_{\text{pre}} \right]^{-1} - R\cdot Q\cdot P \qquad (4.1)$$

where R, Q, and P are small matrices reflecting the impact of the contingency.

- Sparsity techniques have played a major role in advancing contingency screening. These include
  - the compensation method (IMML) for rapidly incorporating the effects of network changes due to contingencies [21]

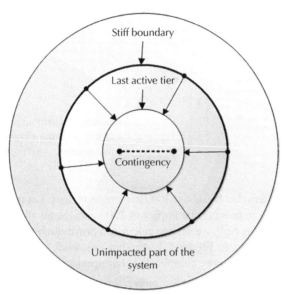

Figure 4.4   The concentric relaxation method.

○ sparse matrix methods enhancing the speed of inversions using factorization methods [19] or partial factorization methods [17]

• Another class of methods that can achieve high speed, reduced memory requirements, and rapid adaptability are localization methods. The concentric relaxation method [22] that introduced this class of solution solves only a small cutoff part of the network in the immediate vicinity of the contingency while treating the remainder as constant voltage. The flows are calculated at second neighborhood level buses called "stiff boundary buses" and compared to their precontingency values. As shown in Figure 4.4, if there is sufficient discrepancy, the subsystem is adaptively enlarged until a sufficient accuracy reached in calculating the boundary flows out of the subsystem. The strength of this method lies in its independence of the local network's solution time from the large system size. The weaknesses of the concentric relaxation method are

1. Unreliable solution algorithm, that is, Gauss–Seidel

2. Possibility of missing severe problems outside the selected solution area

3. Need to solve a number of systems of increasing size

4. Lack of a consistent criteria for selecting the small system buses

More advanced localization methods called bounding methods establish bounding criteria to determine the impacted areas of the network to be solved. The major distinguishing attribute of these methods lies in defining a bounding criterion, which could identify the affected buses anywhere in the system. These methods have been successfully applied to both active power and voltage contingency screening [23].

5. Another method for speeding up the process is to eliminate the unmonitored parts of the network [24]. Elimination is beneficial only while it decreases the computational size. Size can be measured in various ways. It is a function of the reduced network's number of buses, branches, and network matrix factor elements [20]. Elimination of all the unwanted buses may be self-defeating since a number of equivalent branches can be produced reducing matrix sparsity. More advanced network reduction methods provide an optimal network reduction by retaining a minimum number of buses to ensure [25] the sparsity of the reduced matrix factors.

6. New methods [26] have shown that a new layer of prescreening can make the process more efficient by identifying definitely unharmful contingencies. A set of severity measures is employed to reliably identify harmless contingencies from a large set of candidates. It has been shown that this prescreening method is faster than the screening method by a factor of two [27].

4.3.3.2 Contingency Ranking

Once the impact of contingencies is calculated, the severity of contingencies can be ranked. This is done by choosing a performance index in the form of

$$F = \sum f_i \tag{4.2}$$

where f_i is a function denoting overloading conditions.

Examples of F are

$$F = \sum \left| \frac{I_t}{I_{t0}} \right|^2 \tag{4.3}$$

$$F = \sum \left| \frac{(Q_g - Q_{g0})}{Q_{g0}} \right| \tag{4.4}$$

$$F = \sum \left| \frac{(|V_m| - |V_{m0}|)}{|V_{m0}|} \right| \tag{4.5}$$

where I_{t0}, Q_{g0}, and V_{m0} represent the maximum rating values for an electrical current, a generator reactive power, and a bus voltage, respectively.

The performance index does not need to be purely a mathematical formula. For instance, the index might incorporate the largest violation.

The most popular performance index is

$$F = \sum W_i \cdot \left(\frac{x_i}{X_i} \right)^{2m} \tag{4.6}$$

where x is a monitored quantity such as branch flow or voltage, X is its upper limit, W is the weighting factor, and m is a positive integer.

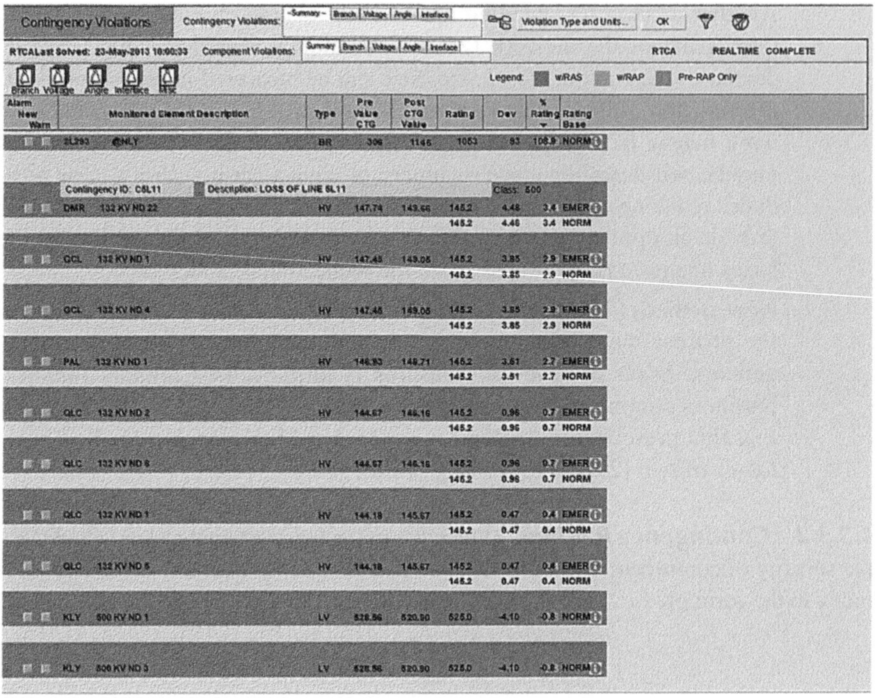

Figure 4.5   A typical contingency screening and ranking display in an EMS.

The choice of $m = 1$ is not very satisfactory since it cannot discriminate between many small violations and a large violation. This phenomenon, which is called masking effect, can be avoided by increasing m.

The choice of f_i is a very important step in contingency severity ranking. The selected f_i should correctly reflect the severity in a broad operating condition range. There have been attempts to develop a method for adjusting the weighting factors in the performance index to maximize the capture rate of the critical contingencies to reduce the masking errors [27].

To perform the ranking, various techniques have been proposed that calculate the sensitivity of the performance index with respect to the outage [28]. These methods perform well when used with a linear power system model such as $P–\delta$ portion of a decoupled load flow. The linearized models for reactive power, so far, have proven to be unreliable due to the lack of capability to incorporate generator VAr limits and ULTC operation.

Figure 4.5 shows a typical display of contingency screening and ranking in an energy management system providing the violations after different contingencies.

4.3.4   Contingency Evaluation

Contingency evaluation is the process of evaluating the selected contingencies. Evaluation should be performed using a power flow. The power flow solution technique should be selected based on which method best satisfies all user requirements.

The evaluation of the critical contingencies will continue until the simulated contingency does not create any violations. It should be noted that all the contingencies ranked lower in the list would not create violations either and the evaluation stops.

Example 4.1

In the three-bus system of Figure 4.6, use a DC power flow to perform active power contingency analysis for three single-branch contingencies.

1. Solve four DC power flows, one for the original system and one for each contingency, and record flows on each line.

2. Use the following performance index to rank the severity of the normal case and the contingency case:

$$S = \sum_{i,j=1}^{3} \left(\frac{P_{ij}}{P_{ij\,max}} \right)^n$$

where P_{ij} is the power flow from bus i to j and $P_{12\,max} = 0.5$, $P_{23\,max} = 1.0$, $P_{13\,max} = 1.2$ and $n = 2$.

Solution

We need to solve DC power flow for the precontingency case, the cases after each contingency:

1. Normal Condition Case

First, we develop the Y matrix $= G + JB$ for the three-bus system

$$Y_{bus} = G + JB = \begin{bmatrix} Y_{11} & Y_{12} & Y_{13} \\ Y_{21} & Y_{22} & Y_{23} \\ Y_{31} & Y_{32} & Y_{33} \end{bmatrix} = - \begin{bmatrix} J15 & -J5 & -J10 \\ -J5 & J25 & -J20 \\ -J10 & -J20 & J30 \end{bmatrix}$$

$$= \begin{bmatrix} -J15 & J5 & J10 \\ J5 & -J25 & J20 \\ J10 & J20 & -J30 \end{bmatrix}$$

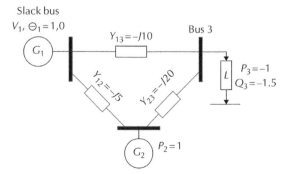

Figure 4.6   Three-bus system.

So, ignoring rows and column 1 associated with Bus 1, we get

$$B = \begin{bmatrix} -25 & 20 \\ 20 & -30 \end{bmatrix}$$

$$P = -B\theta$$

$$\begin{bmatrix} P_2 \\ P_3 \end{bmatrix} = -\begin{bmatrix} -25 & 20 \\ 20 & -30 \end{bmatrix}\begin{bmatrix} \theta_2 \\ \theta_3 \end{bmatrix}$$

$$\begin{bmatrix} \theta_2 \\ \theta_3 \end{bmatrix} = \begin{bmatrix} 25 & -20 \\ -20 & 30 \end{bmatrix}^{-1}\begin{bmatrix} 1 \\ -1.5 \end{bmatrix} = \frac{1}{350}\begin{bmatrix} 30 & 20 \\ 20 & 25 \end{bmatrix}\begin{bmatrix} 1 \\ -1.5 \end{bmatrix} = \begin{bmatrix} 0 \\ -0.05 \end{bmatrix}$$

We will have

$$P_{12} = \frac{(\theta_1 - \theta_2)}{X_{12}} = \frac{(0-0)}{0.01} = 0$$

$$P_{23} = \frac{(\theta_2 - \theta_3)}{X_{23}} = \frac{(0+0.05)}{0.05} = 1$$

$$P_{13} = \frac{(\theta_1 - \theta_3)}{X_{13}} = \frac{(0+0.05)}{0.1} = 0.5$$

$$S_{normal} = \left(\frac{P_{12}}{P_{12\,max}}\right)^2 + \left(\frac{P_{13}}{P_{13\,max}}\right)^2 + \left(\frac{P_{23}}{P_{23\,max}}\right)^2$$

$$S_{normal} = \left(\frac{0}{0.5}\right)^2 + \left(\frac{0.5}{1.2}\right)^2 + \left(\frac{1}{1}\right)^2 = 1.17$$

2. Contingency Line 1–2 Case

In this example, because in the normal condition the flow on line 1–2 is zero, the severity index amount is the same.

3. Contingency Line 1–3 Case

First, we develop the Y matrix $= G + JB$ for the three-bus system:

$$Y_{bus} = G + JB = \begin{bmatrix} Y_{11} & Y_{12} & Y_{13} \\ Y_{21} & Y_{22} & Y_{23} \\ Y_{31} & Y_{32} & Y_{33} \end{bmatrix} = -\begin{bmatrix} J5 & -J5 & 0 \\ -J5 & J25 & -J20 \\ 0 & -J20 & J20 \end{bmatrix}$$

So, ignoring rows and column 1 associated with Bus 1, we get

$$B = \begin{bmatrix} -25 & 20 \\ 20 & -20 \end{bmatrix}$$

$$P = -B\theta$$

$$\begin{bmatrix} P_2 \\ P_3 \end{bmatrix} = -\begin{bmatrix} -25 & 20 \\ 20 & -20 \end{bmatrix}\begin{bmatrix} \theta_2 \\ \theta_3 \end{bmatrix}$$

$$\begin{bmatrix} \theta_2 \\ \theta_3 \end{bmatrix} = \begin{bmatrix} 25 & -20 \\ -20 & 20 \end{bmatrix}^{-1}\begin{bmatrix} 1 \\ -1.5 \end{bmatrix} = \frac{1}{100}\begin{bmatrix} 20 & 20 \\ 20 & 25 \end{bmatrix}\begin{bmatrix} 1 \\ -1.5 \end{bmatrix} = \begin{bmatrix} -0.10 \\ -0.175 \end{bmatrix}$$

We will have

$$P_{12} = \frac{(\theta_1 - \theta_2)}{X_{12}} = \frac{(0 + 0.1)}{(0.2)} = 0.5$$

$$P_{13} = 0$$

$$P_{23} = \frac{(\theta_2 - \theta_3)}{X_{23}} = \frac{(-0.1 + 0.175)}{(0.05)} = 1.5$$

$$S_{\text{Contingency 1-3}} = \left(\frac{0.5}{0.5}\right)^2 + \left(\frac{0}{1.2}\right)^2 + \left(\frac{1.5}{1.0}\right)^2 = 3.25$$

4. Contingency Line 2–3 Case

First, we develop the Y matrix $= G + JB$ for the three-bus system:

$$Y_{\text{bus}} = G + JB = \begin{bmatrix} Y_{11} & Y_{12} & Y_{13} \\ Y_{21} & Y_{22} & Y_{23} \\ Y_{31} & Y_{32} & Y_{33} \end{bmatrix} = -\begin{bmatrix} J15 & -J5 & -J10 \\ -J5 & J5 & 0 \\ -J10 & 0 & J10 \end{bmatrix}$$

So, ignoring rows and column 1 associated with Bus 1, we get

$$B = -\begin{bmatrix} 5 & 0 \\ 0 & 10 \end{bmatrix}$$

$$P = -B\theta$$

$$\begin{bmatrix} P_2 \\ P_3 \end{bmatrix} = -\begin{bmatrix} -5 & 0 \\ 0 & -10 \end{bmatrix}\begin{bmatrix} \theta_2 \\ \theta_3 \end{bmatrix}$$

$$\begin{bmatrix} \theta_2 \\ \theta_3 \end{bmatrix} = \begin{bmatrix} 5 & 0 \\ 0 & 10 \end{bmatrix}^{-1}\begin{bmatrix} 1 \\ -1.5 \end{bmatrix} = \frac{1}{50}\begin{bmatrix} 10 & 0 \\ 0 & 5 \end{bmatrix}\begin{bmatrix} 1 \\ -1.5 \end{bmatrix} = \begin{bmatrix} 0.20 \\ -0.15 \end{bmatrix}$$

We will have

$$P_{12} = \frac{(\theta_1 - \theta_2)}{X_{12}} = \frac{(0 - (0.2))}{(0.2)} = -1.0$$

$$P_{23} = 0$$

$$P_{13} = \frac{(\theta_1 - \theta_3)}{X_{13}} = \frac{(0 + 0.15)}{(0.1)} = 1.5$$

$$S_{Contingency\,2-3} = \left(\frac{-1}{0.5}\right)^2 + \left(\frac{1.5}{1.2}\right)^2 + \left(\frac{0}{1.0}\right)^2 = 5.5625$$

These results indicate that contingency 2–3 represents the worst contingency judging by contingency ranking index selected.

4.3.5 Implementation of Remedial Action Schemes

Through experience and knowledge about their systems, system operators develop RAS for contingencies that normally cause operational infeasibility. These RAS actions normally involve automatic initiation of system or network changes that need to take place after sensing a contingency. For example, an RAS could involve switching in a number of capacitor banks or switching out a number of reactor banks. For each contingency that needs an RAS action, studies need to be conducted to establish the configuration changes. Once the RAS action is obtained, it is armed as an automatic action by SCADA. Operators can arm or disarm a specific RAS action.

QUESTIONS AND PROBLEMS

4.1. What characterizes the static security of a power system?

4.2. System operators develop schemes for contingencies that cause operational infeasibilities. Define and describe these schemes.

4.3. List and describe the three distinct steps in contingency analysis.

4.4. What are some of the contingency types considered by the North East Power Coordinating Council?

4.5. What are sympathetic outages?

4.6. What is a Topology Analyzer and how is it used?

4.7. How does the concentric relaxation method work?

4.8. What is the masking effect?

4.9. In the three-bus system of Figure 4.7, use a DC power flow to perform active power contingency analysis for three single-branch contingencies. Solve four DC power flows, one for the original system and one for each contingency, and record flows on each line.

Use a performance index of $S = \sum_{k=1}^{3} (P_k/P_{k\,max})^n$ to rank the severity of the contingencies in Problem 4.1. Assume $P_{12\,max} = 0.5$, $P_{23\,max} = 1.0$, $P_{13\,max} = 1.2$ and $n = 2$.

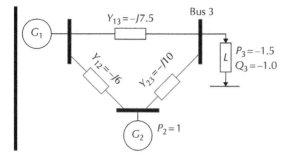

Figure 4.7   Three-bus system.

CHAPTER 5

POWER SYSTEM POSTURING: ANGULAR STABILITY

5.1 OPERATOR'S QUESTION ON POWER SYSTEM POSTURING: ANGULAR STABILITY

The North American Electric Reliability Council (NERC) defines the security of a power system by

1. The presence of acceptable operating conditions before and after a contingency

2. The ability of the system to ride through the contingency and to reach the postcontingency operating condition without becoming unstable

The first condition is called the static security of the system. The second condition, which is a reflection of the system dynamic behavior, is called dynamic security. Dynamic security, in turn, is characterized by two phenomena of angular stability and voltage stability as shown in Figure 5.5.

The operator needs to ensure that the system can ride through any contingency without losing its angular stability glue keeping the generators together. The main question that the operator needs to answer is if the system will be OK after any credible contingency and to the extent that it is not, what actions he or she needs to take to posture the system so that it is operationally feasible.

5.2 PROCESS FOR POWER SYSTEM POSTURING: ANGULAR STABILITY

Angular stability is a fast phenomenon normally taking only a few seconds for the system to become unstable after contingencies. As such operators do not have sufficient to steer time the system away from instability once the contingency occurs. For that reason, the process of posturing the system for angular stability involves developing preventive and corrective measures without any manual interaction from operators after the contingencies.

The preventive measures include creating constraints for the precontingency operating conditions called "nomograms" as shown in Figure 5.1. These

Practical Power System Operation, First Edition. Ebrahim Vaahedi.
© 2014 The Institute of Electrical and Electronics Engineers, Inc. Published 2014 by John Wiley & Sons, Inc.

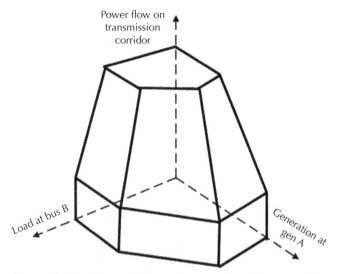

Figure 5.1   Acceptable operating conditions under normal conditions to avoid angular instability under contingencies.

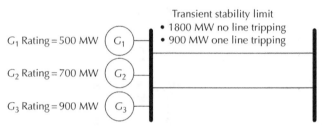

Figure 5.2   A small system demonstrating the impact of generation rejection on angular stability limit.

nomograms restrict precontingency operating conditions to avoid angular instability after contingencies.

Utilities also deploy corrective actions to avoid angular stability. These corrective actions must be designed and implemented under normal operating conditions to avoid angular instability following contingencies. Through experience and knowledge about their systems, utilities develop remedial action schemes (RAS) for contingencies that normally cause operational infeasibility. These RAS normally involve automatic initiation of system or network changes that need to take place after sensing a contingency. For example, an RAS could involve switching in a number of capacitor banks and switching out a number of reactor banks. These remedial action schemes should be in place to provide comfort to the operator to ensure that the system survives angular stability after any credible contingencies.

Let us consider the system of Figure 5.2 assuming that the angular stability limit of this system is 1800 MW with no line tripping and 900 MW with one line

GENERATION SHEDDING DISPLAY (1) REAL TIME GENSHED(2)

CONTINGENCY	TRANSFER TOTAL	SHED TOTAL	SET	GMS 1 2 3 4 5	GMS 6 7 8 9 10	PCN 1 2 3 4	KMO KM0
5L1	+ 500.0	360.0					
5L2	+ 500.0	360.0					
5L3 OR 7	+ 250.0	360.0					
5L4	+ 9.5	0.0					
5L11	+ 530.0	360.0					
5L12	+ 0.0	0.0	●				
5L1 & 2	+ 1000.0	777.7					
5L1 & 3	+ 780.5	360.0					
5L1 & 7	+ 750.0	360.0					
5L2 & 3	+ 780.5	360.0					
5L2 & 7	+ 750.0	360.0					
5L13_1PH		0.0					
5L13	+ 530.0	360.0					
5L11 & 12	TRANSFER	360.0					
5L11 & 13	+ 1057.0	994.0					
5L12 & 13	5L11/12/13	360.0					
2L101	+ 11.4	0.0					
MLS_5CX1		0.0					
MLS_5CX2		0.0	●				
MLS_5CX3		0.0					
CHP_5CX1	+	0.0					
CRK 5CX1	+	0.0					
KDY 5CX1	+	0.0					
KDY 5CX2	+	0.0					
KDY 5CX3	+	0.0	●				
5L40	+ 271.0	0.0					
5L41	+ 555.0	0.0					
5L42	+ 936.0	0.0					
5L43	+ 1.4	0.0	●				
5L44	+ 253.5	0.0					
5L51	+ 135.5	0.0					
5L52	+ 130.6	0.0					
5L51 & 52	+ 266.0	0.0					
C5 L51_52 UF							
5L87	+ 578.0	0.0					
BPA_PACI	BPAPACI RAS (Cct#1)	0.0					
BPA_NW	BPANW RAS (Cct#2)	0.0					
500KV PATH LM-VI							
5L81 & 82	+ 1374.0	551.0	●				

BPANW RAS ARMING LEVEL	☐	GREEN=BPA NORMAL (800 MWS)	RED=BPA ON TV OR 150 MWS					FUTURE
KEEPHILL GENSHED AVAIL	☐	RIGHT CLICK TO CHANGE STATUS		
BRD GENSHED AVAIL	☐			

Figure 5.3   An EMS angular stability displays available to operators at BC Hydro.

tripping. For this system, if we can do generation shedding after the fault, then the angular stability limit with a line tripping increases depending on the unit available for shedding:

Unit G_1 available for shedding: angular stability limit $= 900 + 500 = 1400\,MW$

Unit G_2 available for shedding: angular stability limit $= 900 + 700 = 1600\,MW$

Unit G_3 available for shedding: angular stability limit $= 900 + 900 = 1800\,MW$

The operator on a regular periodic basis needs to know the angular stability limit constraints and the generation shedding armed to achieve that precontingency transfer. For example, if in the example given earlier, unit G_3 is available for shedding, then the precontingency angular stability limit will be 1800 MW provided that unit G_3 is armed for shedding after the tripping of any of the two lines.

In summary, operators are provided with a combination of precontingency nomograms or operating condition constraints as well as corrective actions that must be activated to ensure system angular stability. Operators will need to ensure that the prevailing operating condition remains within the restricted nomograms or constraints. As well, operators may be provided with recommendations for corrective actions or RAS under different contingencies that they need to approve for activation under different contingencies. Figure 5.3 depicts one of the angular stability displays available to operators at BC Hydro. This display provides the operators with corrective actions activated under different contingencies. For example, it shows that after the double contingency 5L1, 5L2, a total of 777.7 MW is shed automatically. The generating units selected for shedding after the double contingency include units 5 and 7 at GMS and the unit at KMO. Figure 5.4 provides another angular stability display indicating the maximum transfer limit on different system corridors and locations. For example, under normal conditions, the display shows that there is a maximum flow limit of 650 MW on F2L129 ARN. Since the limits are calculated offline, the maximum flow limits are also calculated for conditions where a major transmission line is out of service. For example, Figure 3.5 provides the maximum export limit of 850 MW to Alberta (BCH_EXP_AUT) when the two parallel lines of 5L71/5L72 are out of service.

Again going back to our metaphor of operating a car, a good driver, while monitoring the prevailing conditions, needs to ensure that the car speed is under certain limits to ensure that the car survives an accident and reaches a post-accident operating condition without any passengers getting hurt. Some cars are also equipped with air bags (corrective action), which are activated after the incident.

5.3 TECHNOLOGY FOR POWER SYSTEM POSTURING: ANGULAR STABILITY

There are two sets of technologies needed for posturing angular stability:

1. Angular stability assessment
2. Implementation of angular stability limits and remedial action schemes

5.3.1 Angular Stability Assessment

The security of a power system is characterized by

1. The presence of acceptable operating conditions before and after a contingency
2. The ability of the system to ride through the contingency and to reach the post-contingency operating condition without becoming unstable

TSA Limits

GROUP: OO717(TAU) Template: TAU 5L71-72

LIMIT NAME	PRESENT	LIMIT
BCH EXP TAU	306.9	850.0
BCH IMP TAU	-306.9	691.1
EMRG EXP TAU	306.9	850.0
EMRG IMP TAU	-306.9	691.1
F2L293	-124.8	

GROUP: OO718(BPA) Template: BPA 5L71 72

LIMIT NAME	PRESENT	LIMIT
ING EXP CUS	-468.0	2000.0
ING IMP CUS	468.0	2000.0
BCH EXP BPA	-523.4	2400.0
BCH IMP BPA	523.4	2000.0
BUT EQUIV SC	0.0	0.0
BCH LOAD	6300.8	

GROUP: OO713(PEACE) Template: PS 5L12 45L

LIMIT NAME	PRESENT	LIMIT
GMS PCN OUT	1467.8	3265.3
GMS PCN SWT	1467.8	
GMSTMW	1158.3	
GMS ING ANG	14.3	
BCH EXP ALC	-246.8	0.0
BCH IMP ALC	246.8	

GROUP: OO71T(SYS) Template: BCH SYS

LIMIT NAME	PRESENT	LIMIT
TTL IMP	463.3	2994.6

GROUP: OO734(SEL) Template: SEL NORM

LIMIT NAME	PRESENT	LIMIT
KCL OUTPUT	524.2	
ALHTMW	94.2	
SEVTMW	739.0	
KCL SEV OUT	1263.2	
SEL GEN	1903.7	
KCL ALM	618.4	
SEL GEN S3 4	1543.7	
ALH SEV	833.2	
SEL GEN S1 2	1524.7	
SSEL T1 2 3	1312.5	
SSEL T MVA	1832.1	
F2L295 2L299	589.7	
F2L293 NLY	125.3	

GROUP: OO714(BR) Template: BR 2L90

LIMIT NAME	PRESENT	LIMIT
BRT OUTPUT	87.8	
BR UH WAH	217.0	736.0
BR UH	217.0	
F2L90 BRT	0.0	
F3L2 BRT	54.0	
F2L2 RUT RBW	77.8	
ROS T1 MW	177.1	470.0
F2L1 PEM RBW	62.9	
F2L9 CKY	216.2	
F2L13 CKY	191.6	
FBRT T4	38.9	

GROUP: OO734(SI) Template: SI 5L71 72

LIMIT NAME	PRESENT	LIMIT
MCA REV OUT	1091.0	
MCA OUTPUT	0.0	
REV OUTPUT	1091.0	2518.1
COLUMBIA GEN	2903.8	
F5L81 82	1276.9	
F5L81 87	1450.6	
F5L82 87	1131.3	
F5L81 82 41	1743.4	
F5L91 96	1466.8	2434.1
F2L112 NLY	-55.4	
F2L112 BDY	55.4	
F2L293 NLY	125.3	
FBC INJ	589.2	1280.3
SI PCN OUT	2558.8	

GROUP: OO741(VI5) Template: VI5 NORM

LIMIT NAME	PRESENT	LIMIT
LM2VI	852.8	
F2L129 ARN	27.1	590.0
VI MR GEN	227.3	
LM2VI500	825.6	

GROUP: OO741(VI2) Template: VI2 NORM

LIMIT NAME	PRESENT	LIMIT
LM2VI230	27.1	
F2L123 28DMR	591.8	1465.0

Figure 5.4 An EMS angular stability displays available to operators at BC Hydro.

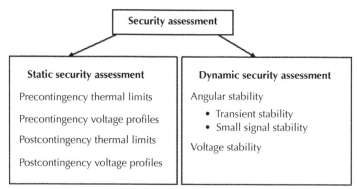

Figure 5.5   Security assessment of the system.

The first condition is called the static security of the system. The second condition, which is a reflection of the system dynamic behavior, is called the dynamic security of the system.

To assess the security of a system, both of the conditions should be checked. To check for the presence of the pre- and postcontingency operating condition, the following limits should be checked:

1. Thermal limits
2. Voltage profile limits
3. Postcontingency thermal limits
4. Postcontingency voltage profile limits

The checking against the precontingency thermal limits and voltage profile limits involves the comparison of the precontingency operating condition against these limits. The checking against postcontingency thermal limits as well as voltage profile limits is more involved as it requires the simulation of different contingencies to determine the postcontingency system conditions that should be checked against these limits. This has been explained in detail in Chapter 4 dealing with contingency analysis.

To assess the dynamic security of a system, the dynamic system performance has to be derived from the two viewpoints, angular stability and voltage stability, as shown in Figure 5.5. To evaluate the dynamic response of the system after a contingency, therefore, the two limits, angular and voltage stability, need to be calculated. Traditionally, these limits have been derived using elaborate offline computer simulations. Both of these limits are translated into system parameters such as maximum transfer limits (precontingency flows on major interfaces), maximum generation at major plants, or remedial actions such as generation shed or load shed.

In Chapter 4 dealing with contingency analysis, it is shown how postcontingency assessment for thermal and voltage is made to identify violations. In this section, how the dynamic response of the system following a contingency is assessed and the dynamic security limits are derived and implemented to ensure the security of the systems is shown.

5.3.2 Power System Stability

Power system stability can be broadly defined as

The capability of the system to remain in a state of operating equilibrium under normal operating conditions and to regain an acceptable state of equilibrium after being subjected to a disturbance.

Power system stability can be classified into the following three categories based on the time period of the phenomenon under the study:

- Short term or transient: 0–10 seconds
- Mid term: 10 seconds to a few minutes
- Long term: a few minutes to tens of minutes

5.3.3 Angular Stability

The angular stability of a power system can be defined as

The capability of the system to ride through contingencies without breaking the synchronizing glue between the machines and reaching a new steady-state condition with sufficient damping.

The angular stability of a power system can be divided into two types:

1. Small-signal stability
2. Transient stability

Small-signal stability is defined as *the capability of the system to withstand small disturbances with sufficient damping.*

Transient stability is defined as *the capability of the system to withstand contingencies without machines losing their synchronism.*

In the following subsection, the techniques for assessing the transient and small-signal stability of a power system are described.

5.3.4 Transient Stability

The equation of motion for a machine can be written as

$$P_{mech} - P_{elec} = \left(\frac{2H}{\omega_0} \right) \frac{d^2\delta}{dt^2} \tag{5.1}$$

where, P_{mech}, is the mechanical input power to the machine; P_{elec}, is the electrical power output of the machine; H is the inertia constant of the machine in MW. S/MVA, which is defined as

$$H = \frac{0.5I\omega_0^2}{\text{Rated MVA}}$$

where I is the machine inertia, ω_0 is the synchronous speed, which is related to synchronous frequency f_0 (60 Hz) by

$$\omega_0 = 2 \cdot \pi \cdot f_0 \tag{5.2}$$

δ is the electrical angle, which is related to the mechanical angle by

$$P_{mech} - \delta_{elec} = \delta_{mech}\left(\frac{N}{2}\right) \tag{5.3}$$

where N is the number of the generator poles.

Please note that this equation assumes that the speed variations are small and hence the per-unit value of torque is equal to power.

5.3.5   Small System

In order to make the concepts more clear, the single-machine infinite bus system of Figure 5.6 is analyzed.

By representing the machine by a simple constant-voltage-behind a reactance model, the model shown in Figure 5.7 is obtained, in which the following variables have been assigned:

V_f: machine field voltage

E, δ: machine internal voltage and phase angle induced by field voltage

V_t, δ_t: terminal voltage and phase angle

$V_{inf}, 0$: infinite bus voltage with the reference angle of zero

X':   machine internal reactance

X_l: transmission line reactance

Using this model, the electric power out of the machine can be derived by

$$P_e = \frac{E \cdot V_{inf}}{X' + X_l}\sin(\delta) \tag{5.4}$$

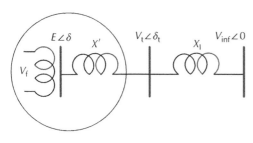

Figure 5.6   Single-machine infinite bus system.

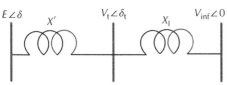

Figure 5.7   Equivalent infinite bus system.

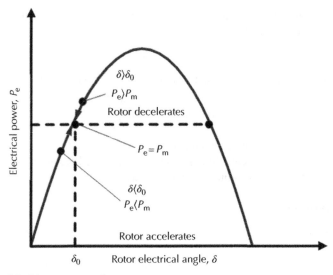

Figure 5.8   Machine power angle curve.

The synchronous machine power angle curve can be obtained using Equation 5.4. This curve is given in Figure 5.8. As this figure shows, the value of the machine angle is determined by the value of P_{mech}. The maximum angle is 90°, corresponding to P_{max}. Equation 5.1 can be written as

$$\frac{2H}{\omega_0} \cdot \frac{d^2\delta}{dt^2} = P_{\text{m}} - \frac{E \cdot V_{\text{inf}}}{\left(x' + x_1\right)} \sin(\delta) \tag{5.5}$$

It is noted that in this analysis it is assumed that E and P_{mech} are constant (no AVR and governor effect).

To analyze the system, the second-order equation would have to be integrated with time to produce the variation in angle, speed, and other related parameters. To simplify the equation, it is converted to two first-order differential equations by defining another variable:

$$\frac{d\delta}{dt} = \omega \tag{5.6}$$

Equation 5.5 can now be written as

$$\frac{d\omega}{dt} = \frac{d^2\delta}{dt^2} = \frac{\omega_0}{2H}(P_{\text{m}} - P_{\text{e}}) \tag{5.7}$$

Writing Equations 5.6 and 5.7 in a matrix form results in

$$\begin{bmatrix} \dfrac{d\delta}{dt} \\[2mm] \dfrac{d\omega}{dt} \end{bmatrix} = \begin{bmatrix} 0 & 1 \\ 0 & 0 \end{bmatrix} \begin{bmatrix} \delta \\ \omega \end{bmatrix} + \begin{bmatrix} 0 \\[2mm] \dfrac{P_{\text{m}}}{J} - \dfrac{1}{J}\dfrac{E \cdot V_{\text{inf}}}{\left(x' + x_2\right)} \sin(\delta) \end{bmatrix} \tag{5.8}$$

where

$$J = \frac{2H}{\omega_0} \qquad (5.9)$$

In order to obtain variations of angle and speed, the aforementioned first-order differential equation has to be integrated in time. It is noted that the value of P_e changes in different stages of simulation. In the perfect stage, $P_e = P_{mech}$. After the fault inception, P_e drops to a very low value (can be assumed zero). After the fault recovery, the value of P_e should also be adjusted for the change in the line impedance if it changes (loss of a circuit due to the fault).

5.3.6 Integration Methods

Power system stability can be represented by a set of differential equations in the form

$$x' = f(x,t) \qquad (5.10)$$

To integrate this set of differential equations, two types of integration methods are available, namely, explicit methods and implicit methods. These methods are described next.

5.3.6.1 *Explicit Methods*   In the explicit methods, the value of x at time t is obtained from the knowledge of x from the previous time step. In other words, the value of x at the $(n+1)$ step is obtained from the value at the (n) step. These methods are simple to implement, but the problem with these methods is that the integration time step is limited by the shortest time constant.

Euler Method   The objective here is to obtain the value of $X(t+h)$ assuming that the values of $X(t)$ and $X'(t)$ are known. The Taylor expansion for $X(t+h)$ gives

$$X(t+h) = X(t) + hX'(t) + 0.5(h^2)X''(t) + \cdots \qquad (5.11)$$

Ignoring second-order terms of those of higher order results in

$$X(t+h) = X(t) + hX'(t) \qquad (5.12)$$

This equation is the basis of the Euler method. This is also shown graphically in Figure 5.9. This method only considers the first two terms of the Taylor series, and therefore the error is in the order of h^2.

Example 5.1
Integrate $X' = t^2$ for one time step with $h = 0.1$, $X_0 = 1$ and $t_0 = 1$.

$$X'(t_0) = t_0^2 = 1$$

$$X(t_0 + h) = X_0 + ht_0^2 = 1 + 0.1 \times 1 = 1.1$$

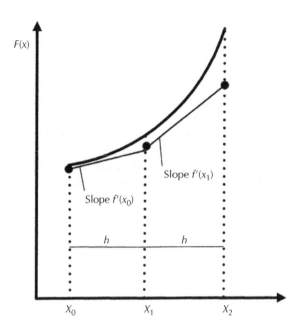

Figure 5.9   Euler method.

Modified Euler Method   The modified Euler method is obtained by using an average slope in Equation 5.8. The average slope is given by

$$X'_{av} = 0.5\left[\left(X'(t) + X'(t+h)\right)\right] \tag{5.13}$$

Example 5.2
Integrate $X' = t^2$ using modified Euler method for one time step with $h = 0.1$ and $t_0 = 1$ and $X_0 = 1$

Solution

$$X'_{av} = 0.5\left[\left(X'(t) + X'(t+h)\right)\right] = 0.5\left(t_0^2 + (t_0 + 0.1)^2\right)$$

$$X'_{av} = 0.5(1 + 1.21) = 1.105$$

$$X(t+h) = X(t) + h * X'_{av}(t) = 1 + 1.105 \times 0.1 = 1.1105$$

Runge–Kutta Method   This method performs the integration in four steps:

$$x_1 = hX'(t,X)$$

$$x_2 = hX'\left(t + \frac{h}{2}, \frac{x_1}{2} + X\right)$$

$$x_3 = hX'\left(t + \frac{h}{2}, \frac{x_2}{2} + X\right) \tag{5.14}$$

$$x_4 = hX'(t + h, x_3 + X)$$

$$X(t+h) = X(t) + \left(\frac{1}{6}\right)(x_1 + 2x_2 + 2x_3 + x_4)$$

This method is more accurate than the Euler method but at the expense of more computing time. It can be shown that the error in this method is of the order of h^5.

Example 5.3

Solve $X' = t^2$ using the Runge–Kutta method with $h=0.1$, $t_0=1$ and $X_0=1$

$$x_1 = hX'(t,X) = 0.1 \times t_0^2 = 0.1$$

$$x_2 = h \cdot X'\left(t+\frac{h}{2}, X+\frac{x_1}{2}\right) = 0.1(X_0 + 0.05)2 = 0.1 \times 1.052 = 0.11025$$

$$x_3 = h \cdot X'\left(t+\frac{h}{2}, X+\frac{x_2}{2}\right) = 0.1\left(X_0 + \frac{0.11025}{2}\right)2 = 0.1111$$

$$x_4 = h \cdot X'(t+h, X+x_3) = 0.1 \times (1+0.0556)2 = 0.1114$$

$$X(t+h) = X(t) + \left(\frac{1}{6}\right)(x_1 + 2x_2 + 2x_3 + x_4)$$

$$X(0.1) = 1 + \frac{1}{6}(0.1 + 2 \times 0.11025 + 2 \times 0.1111 + 0.1114) = 1.1090$$

5.3.6.2 Implicit Methods   In the implicit methods, the value of x at time t is obtained from the value of x in the previous time steps and the value at time t. Integrating Equation 5.7 gives

$$X_{n+1} = X_n + \int_{t_n}^{t_{n+1}} f(x,t) \cdot dt \tag{5.15}$$

The simplest implicit method is the trapezoidal rule. In this method, the value of the integral is obtained by interpolation as shown in Figure 5.10, which simplifies Equation 5.15 to

$$X_{n+1} = X_n + \frac{\Delta t}{2}\left[f(x_n,t_n) + f(x_{n+1},t_{n+1})\right] \tag{5.16}$$

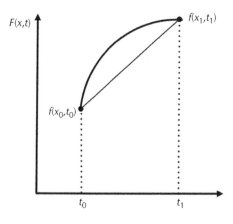

Figure 5.10   Trapezoidal method of integration.

This equation would have to be solved iteratively as X_{n+1} appears on both sides of the equation. The main advantage of this method is that with larger time steps the numerical stability does not occur. With larger time steps, the high-frequency modes are filtered out and the solution for slower modes is accurate.

5.3.7 Equal-Area Criteria Method

For the simple single-machine infinite bus system of Figure 5.5 and Figure 5.6, Equation 5.1 is written in the form

$$\frac{d^2\delta}{dt^2} = \frac{\omega_0}{2H}(P_m - P_e)$$

(5.17)

If both sides of this equation are multiplied by $2d\delta/dt$,

$$2\frac{d\delta}{dt} \cdot \frac{d^2\delta}{dt^2} = \frac{W_0(P_m - P_e)}{H}\frac{d\delta}{dt}$$

(5.18)

This gives

$$\frac{d}{dt}\left[\frac{d\delta}{dt}\right]^2 = \frac{\omega_0(P_m - P_e)}{H} \cdot \frac{d\delta}{dt}$$

(5.19)

Integrating both sides of this equation gives

$$\left[\frac{d\delta}{dt}\right]^2 = \int \omega_0 \frac{(P_m - P_e)}{H} d\delta$$

(5.20)

This is a very important equation. It states that if the system remains stable and the angle reaches a maximum value, the speed should become zero, meaning that Equation 5.20 can be written as

$$\int_{\delta_0}^{\delta_m} \frac{\omega_0}{H}(P_m - P_e)d\delta = 0$$

(5.21)

where δ_0 and δ_m are the initial and maximum values for the rotor angle. Figure 5.11 and Figure 5.12 clarify Equation 5.21. There are three curves in these figures to reflect the machine's electrical power during prefault period, fault-on period, and postfault period. Assuming that the fault is cleared when the angle is equal to δ_c, then area 1 reflects the accelerating period when the machine's kinetic energy increases. This area is

$$Area\ 1 = \int_{\delta_0}^{\delta_{cl}} (P_m - P_e)d\delta$$

(5.22)

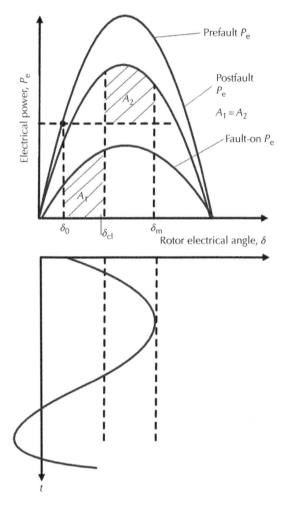

Figure 5.11   System stable after a fault cleared at t_{cl}.

Area 2 belongs to machine deceleration when the fault is cleared up to the maximum angle of δ_m. Area 2 can be calculated by

$$Area\ 2 = \int_{\delta_{cl}}^{\delta_m} (P_e - P_m)d\delta \qquad (5.23)$$

The equal area states that these two areas should be equal for the system to be stable. The criteria can be used to assess system stability by comparing area 1 with the maximum value of area 2. It can also be used to obtain the maximum value for fault-clearing angle or $\delta_{critical}$ for which the system can be stable.

From the calculated $\delta_{critical}$, the maximum fault-clearing time called $t_{critical}$ can be obtained. This is done by assuming that during the fault, P_e is zero, changing Equation 5.1 to

$$w_0 \cdot \frac{P_m}{2H} = \frac{d^2\delta}{dt^2} \qquad (5.24)$$

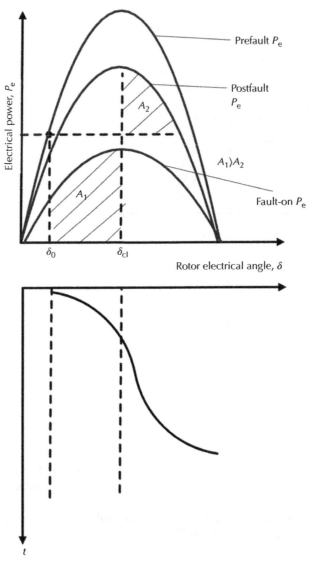

Figure 5.12   System unstable after a fault cleared at t_{cl}.

Integrating Equation 5.24 between δ_0 and δ_c results in

$$\delta_c = \delta_0 + \frac{w_0 \cdot P_m M \cdot t_c^2}{4H} \tag{5.25}$$

giving a fault clearing-time value of

$$t_c = \sqrt{\frac{(\delta_c - \delta_0)}{(4H / w_0 \cdot P_m)}}$$

(5.26)

Example 5.4

For the following system (Figure 5.13),

1. Write the equation of motion in the form of two first-order differential equations.
2. Find the steady-state conditions and E, δ.
3. Using equal-area criteria, calculate the critical angle and critical clearing time.
4. Write the differential equations for the three periodsprefault, fault-on, and postfault.

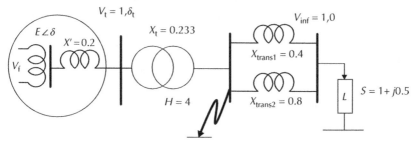

Figure 5.13   System diagram in Example 5.4.

Solution

Generally, a dynamic simulation involves solving the following second-order differential equations in the three periods—prefault period, fault period, and post-fault period—where t_c is the fault-clearing time:

$$\frac{d^2\delta}{dt^2} = \frac{\omega_0}{2H}(P_m - P_e)$$

First, we change the second-order differential equations into two first-order differential equations by defining a second variable ω, which represents the derivative of δ_1:

$$\frac{d\delta_1}{dt} = \omega$$

$$\frac{d\omega}{dt} = \frac{\omega_0}{2H}(P_m - P_e) = \frac{377}{8}(1.0 - A\sin\delta_1)$$

where

$$\omega_0 = 2\pi f = 2\pi 60 \approx 377$$

$$P_m = 1.0$$

$$H = 4$$

$$P_e = A\sin\delta_1$$

1. To get the steady-state condition, we first consider the steady-state network as shown (Figure 5.14):

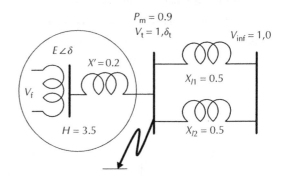

Figure 5.14   System diagram of Example 5.4 and the steady-state prefault condition.

For the equivalent system, we can write the following equation

$$P_m = P_e = \frac{V_t V_{inf}}{(X_{equivalent})}\sin(\delta_t - \delta_{inf})$$

where, $X_{equivalent}$ is the equivalent impedance between V_t and V_{inf} calculated next:

$$X_{equivalent} = X_t + \frac{1}{\left(\dfrac{1}{X_{l1}} + \dfrac{1}{X_{l2}}\right)}$$

$$X_{equivalent} = 0.233 + \frac{1}{\left(\dfrac{1}{0.8} + \dfrac{1}{0.4}\right)} = 0.5$$

$$P_m = P_e = 1 = \frac{V_t V_{inf}}{(X_{equivalent})}\sin(\delta_t - \delta_{inf}) = \frac{\sin(\delta_t - 0)}{0.5}$$

$$\delta_t = 30$$

$$E = V_t + IX'_d$$

$$I = \frac{(V_t - V_{inf})}{(J0.5)}$$

$$E = V_t + J0.3(IX'_d)$$

$$E = 0.8526 + j0.7232$$

2. To use the equal-area criteria, the network equivalents before the fault, during the fault, and after the fault are obtained as shown:

Before the fault (Figure 5.15),

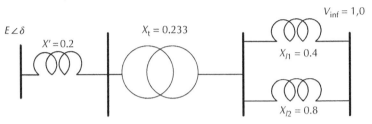

Figure 5.15   Network equivalent of Example 5.4 during the prefault condition.

$$P_e = \frac{1.1180 \times 1}{0.6997} \sin \delta = 1.5980 \sin \delta$$

During the fault-on period, the voltage at the faulted bus is zero and hence P_e becomes zero (Figure 5.16):

$$P_e = EV_{\text{faulted bus}} \sin(\delta - \delta_{\text{faulted bus}}) = 0$$
$$P_e = 0$$

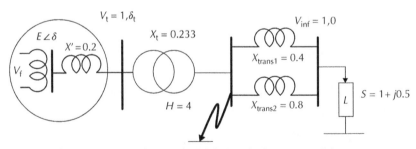

Figure 5.16   System diagram of Example 5.4 during the fault-on condition.

After the fault recovery, the faulted line is lost, resulting in Figure 5.17 and the Equal Area Criteria diagram shown in (Figure 5.18):

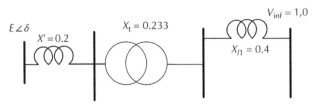

Figure 5.17   Network equivalent of Example 5.4 during the fault-on condition.

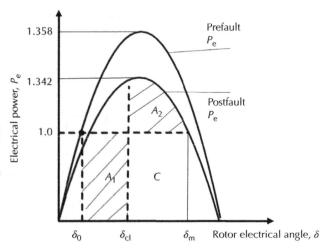

Figure 5.18   Equal area criteria diagram for Example 5.4.

$$P_e = \frac{1.1180 \times 1}{0.8330} \sin\delta = 1.3422 \sin\delta$$

$$1.3422 \sin\delta_m = 0.9 \rightarrow \pi - \delta_m = 2.4066 \text{ rad}$$

$$1.5980 \sin\delta_0 = 0.9 \rightarrow \delta_0 = 0.5983 \text{ rad}$$

$$A = A_2 \rightarrow A_1 + C = A_2 + C$$

$$0.9(2.4066 - 0.5983) = \int_{\delta_c}^{2.4066} 1.1024 \sin\delta \; d\delta$$

$$1.6275 = -1.3422 \cos\delta]_{\delta_c}^{2.4066} \rightarrow 1.627 = (-1.3422 \cos 2.4066) - (-1.3422 \cos\delta_c)$$

$$0.6318 = 1.3422 \cos\delta_c \rightarrow \delta_c = 1.0807 \text{ rad}$$

$$t_c = \sqrt{\frac{(\delta_c - \delta_0)4H}{\omega_0 P_m}} = \sqrt{\frac{(1.0807 - 0.5983)4 \times 4}{2\pi 60 \times 0.9}}$$

$$t_c = 0.1508 \text{ seconds}$$

3. The equations of motion for the three periods prefault, fault-on, and postfault are:

$$\frac{d^2\delta}{dt^2} = \frac{\omega_0}{2H}(P_m - P_e)$$

where $H=4$, $P_m=1.0$, and $P_e = A \cdot \sin\delta$ for which the value of A is 1.598 during the prefault period, zero during the fault-on period, and 1.342 during the postfault period.

5.3.8 Models for Other Components

In the equation, the values of P_m and P_e (or more correctly, T_m and T_e, the mechanical and electrical torques) themselves are functions of other system components that have been so far ignored in this write-up. P_m, the mechanical power delivered to the generator, is the output of the turbine governor system. Turbines produce the mechanical power, which is regulated by the governor system based on the system frequency. Similarly, P_e is a function of generator internal voltage that is dependent on the system excitation as well as its regulating loop, which is automatic voltage regulator (AVR). The excitation system produces the field voltage for the machines and is regulated by AVR, which attempts to retain the generator's terminal voltage at a set value. Excitation systems are also equipped with stabilizers for improving system stability. There are also other components in the system such as motors, HVDC systems, SVCs, and loads, that need to be modeled.

To assess system stability, all other system components such as those mentioned earlier should be included in the modeling. The differential equations representing other components should be added to Equation 5.8.

5.3.9 Multimachine System

For assessing the behavior of a multimachine system, the same procedure used for a single-machine infinite bus is used. The differential equations used to represent each machine are put together to represent the system. In simple terms, the overall system equations including the differential equations for the devices and the algebraic equation relating current injections from each device to the network currents can be written as [29,30]

$$x' = F(x,V) \tag{5.27}$$

$$I(x,V) = Y_N \cdot V \tag{5.28}$$

where x is the state vector of the system, I and V are bus voltage and current injection vectors.

There are two approaches to solve these equations, namely, the sequential approach and the simultaneous approach [29].

In the sequential approach, the following steps are followed:

1. In the steady-state condition, set all the values set x' equal to zero and get all the steady-state values for x, V, and I.

2. After a disturbance, the algebraic Equations 5.28 are first solved to obtain I and V, assuming that x cannot change immediately.

3. New values for V are used in Equation 5.27, and it is integrated to obtain the value of x at the next time step.

4. If simulation time is not reached, go back to 2.

In the simultaneous approach, the following steps are followed:

1. In the steady-state condition, set all the values set x' equal to zero and get all the steady-state values for x, V, and I.

2. After a disturbance, solve the differential and algebraic Equations 5.27 and 5.28 together simultaneously. This method is especially suitable for implicit integration methods, which perform the integration using an iterative method. Equation 5.28 is converted to the following algebraic equation, which can then be solved simultaneously with Equation 5.27:

$$X_{n+1} = X_n + f(V_n, X_n, X_{n+1}, V_{n+1}) \tag{5.29}$$

3. If simulation time is not reached, go back to 2.

5.3.10 Small-Signal Stability

As described earlier, the angular stability of a power system can be divided into two types—transient stability and small-signal stability, where the latter is defined as *the capability of the system to withstand small disturbances with sufficient damping*.

To obtain the equations to assess small-signal stability, the rule of linearization or small change is introduced. Assume A is a function of different variables δ, ω, etc. The rule of small change says that the variation in A is related to the variations in other variables as follows:

$$A = f(\delta, \omega, \ldots) \tag{5.30}$$

$$\Delta A = \frac{df}{d\delta} \Delta\delta + \frac{df}{d\omega} \Delta\omega + \cdots \tag{5.31}$$

If this rule is applied to the single-machine infinite bus system (Eq. 5.6) knowing that P_m and V_{inf} have constant values, then the following equation is obtained:

$$d \begin{bmatrix} \Delta\delta \\ \Delta\omega \end{bmatrix} = \begin{bmatrix} 0 & 1 \\ -\dfrac{E \cdot V_{inf}}{J(x' + x_1)} \cos\delta & 0 \end{bmatrix} \begin{bmatrix} \Delta\delta \\ \Delta\omega \end{bmatrix} \tag{5.32}$$

This equation is in the form of $X' = AX$, and it is easier to solve this than solving the original nonlinear equation. An easy method for solving this equation is through frequency methods. Eigenvalue and Eigenvectors can be derived for the matrix A. Eigenvalues represent different modes of the system. Eigenvectors reflect the participation of each mode in the state variables. A positive eigenvalue represents an unstable mode. The complex-conjugate modes having the same real part and reverse-sign complex part represent an oscillatory mode. The frequency of the complex mode and its damping are shown in Figure 5.19.

NERC has a specific standard on the minimum allowable system damping. As a rule, the system should not have any unstable modes, and the damping of oscillatory modes should be more than a specific value (e.g., 0.03).

Figure 5.20 demonstrates the derivation of oscillation damping from a system time response. By connecting the maximum or minimum points, an exponential

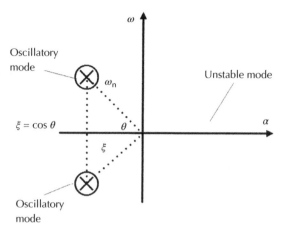

Figure 5.19   Frequency and damping of the conjugate mode.

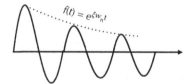

Figure 5.20   Measuring damping from oscillations.

curve is obtained in the form of $f(t) = e^{-\xi w_n t}$. From this curve, the damping value ξ can be calculated, considering that w_n is available by measuring the frequency or the number of oscillations in one second.

The major shortcoming of the conventional eigenvalue methods is that because of computational difficulties it can handle only up to 500 states [1]. Lately, methods have been proposed that calculate only a selected number of eigenvalues [31]. These methods can handle large-scale power system models.

5.3.11   Angular Stability Limit Derivation

To formulate the angular stability results of a power system, the idea is to select a number of system stability features and translate system stability results to these features [32]. By mapping the stability results to operating parameters, an operating region is obtained within which the system is stable. The system stability features that are also called "operating parameters," "transfer limits," or "flow gates" are normally selected from prominent system features, which play a major role in system stability such as

- The total power flow on a branch
- The total power flow on an interface or a corridor
- The total generation at a plant
- The total load in one area
- System voltage at a specific system location

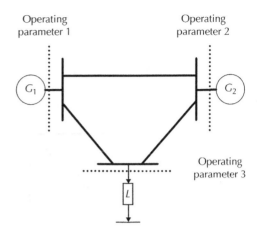

Figure 5.21 Small system with
defined operating parameters.

To formulate operating guidelines that ensure the secure operation of the system, a large number of offline analytical power system simulations are conducted to examine the stability of the system in relation to each "operating parameter." In other words, we need to establish the "operating parameter" values at which the system becomes unstable.

As an example, assume that we want to obtain the system conditions for which the system shown in Figure 5.21 is stable during the normal conditions and following the loss of any transmission system. We select three operating parameters:

1. Operating parameter 1: total generation at bus 1
2. Operating parameter 2: total generation at bus 2
3. Operating parameter 3: total load at bus 3

To develop the operating region and formulate operating guidelines that ensure the secure operation of the system, a large number of offline analytical power system simulations are conducted to stress one "operating parameter" at a time to get its maximum value, assuming that other operating parameters have constant values. In this case, we assume that there are correlations between different values for operating parameters, that is, the maximum value of one operating parameter depends on the values that other operating parameters have. Let us assume that after a number of steady-state and transient stability studies, we get the following nomogram reflecting the maximum transient stability limits (Figure 5.22).

Operators need to ensure that they operate within the sphere of the nomogram. To do that, we need to formulate this nomogram into the energy management system and ensure that the system operating condition remains within it.

A typical sequence of studies conducted to identify system stability is

1. Choose the worst operating condition for the period to study. This includes the total system demand, generation pattern, and transactions with adjacent utilities.

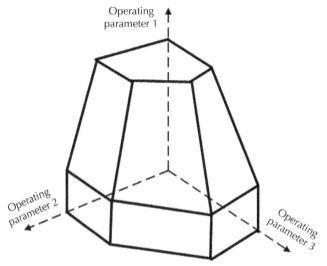

Figure 5.22   Stable operating region for system stability.

2. Assess precontingency violations using a load flow solution.

3. Assess the small-signal stability of the system.

4. Determine the stability limits by increasing the operating parameters until the system becomes unstable following one of the most severe likely contingencies.

In the list, item 3 should be repeated until two cases are obtained: one unstable, one stable. The value of the operating parameter is determined, and the limit will be the value determined minus a confidence margin established by the utility.

5.4   IMPLEMENTATION OF ANGULAR STABILITY LIMITS

Once the system transient stability is obtained in terms of "operating parameter limits" from offline studies, these limits are then formulated in decision tables in the energy management system [32]. For example, in the small example earlier, the nomogram given in Figure 5.16 needs to be entered into the energy management system as allowable domain of system operation. At any point of time, the energy management system calculates from the measured data the values of the operating parameters and compares them against the limits. It warns the operator in case an operating parameter violated. Furthermore, if the utility also deploys corrective actions to avoid transient stability, tables with recommended corrective actions for different contingencies are provided to operators to arm the recommended corrective actions for activation after contingencies. An example of such a table is provided in Figure 5.3.

QUESTIONS AND PROBLEMS

5.1. Define power system security and power system dynamic security.

5.2. Define power system stability and its time frame classifications.

5.3. Describe a typical sequence of studies that need to be conducted to identify system stability limits.

5.4. Solve the following differential equations using the Euler method, with $y(0)=0$ and a time step of 0.1 with two iterations.

    a. $\dfrac{dy}{dx} = 2x^2$

    b. $\dfrac{dy}{dx} = 0.1\left(x^2 + \dfrac{3}{2}x + 4\right)$

    c. $\dfrac{dy}{dx} = x^2$

5.5. Solve the following differential equation using the Runge–Kutta method for two time steps of 0.1 with two iterations.

$$\frac{dy}{dx} = 2x^2$$

5.6. For the following second differential equation, compare the results:

$$\frac{d^2x}{dt^2} = -\sin(x)$$

The initial conditions are $t=0$. Let $h = \Delta t = 0.1$

$$x(0) = 0.5\,\text{rad}$$

$$\frac{dx(0)}{dt} = 0$$

    a. Solve with Euler's method with one iteration.

    b. Solve with modified Euler method with one iteration.

    c. Solve with trapezoidal method with one iteration.

5.7. Solve the problem of Question 9.6 using the Runge–Kutta method for one iteration by hand.

5.8. Write a MATLAB code to solve the aforementioned equation.

5.9. For the following system, first establish the steady-state conditions and E, δ. Then, using the equal-area criteria, calculate the critical angle and fault-clearing time (Figure 5.23).

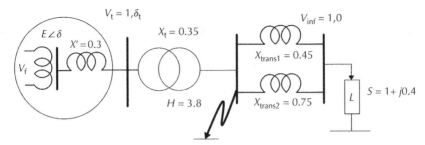

Figure 5.23   System diagram for Question 5.9.

With the following values

- $X_{trans1} = 0.45$
- $X_{trans2} = 0.75$
- $X_t = 0.35$
- $V_t = 1$
- $H = 3.8\,\text{MW second/MVA}$
- $X' = 0.3$
- $S = 1.0 + j0.4$
- $V_{inf} = 1\angle 0$

5.10. Write the equations of motion for the three periods: prefault, fault on, and postfault. Integrate the equations of motion using the Euler integration method with a time step of 0.0001 to obtain the critical clearing time assuming the following periods:

Prefault : $t \le 1$

Fault $-$ on : $1 \le t \le 1 + t_{cl}$

Postfault : $t \ge 1 + t_{cl}$

CHAPTER 6

POWER SYSTEM POSTURING: VOLTAGE STABILITY

6.1 OPERATOR'S QUESTION ON POWER SYSTEM POSTURING: VOLTAGE STABILITY

The security of a power system is characterized by

1. The presence of acceptable operating conditions before and after a contingency
2. The ability of the system to ride through the contingency and to reach the post-contingency operating condition without becoming unstable

The operator needs to ensure that the system has the ability to ride through the contingency and reach the postcontingency operating condition without becoming unstable. The two phenomena impacting the system ability to ride through contingencies are angular stability and voltage stability. Power system voltage stability is defined [31] as

the ability of the system to maintain steady acceptable voltages at all the buses in the system at normal operating conditions and after being subjected to a disturbance.

The operator needs to ensure that the system can ride through any contingency without becoming voltage-unstable; losing stability when voltages uncontrollably decrease due to a disturbance such as an outage, increase in load, decrease in generation, or generally any changes in the operating condition.

The main question that the operator needs to answer is if the system will be voltage-stable after any credible contingency and to the extent that it is not, what actions he or she needs to take to posture the system so that it is operationally feasible.

6.2 PROCESS FOR POWER SYSTEM POSTURING: VOLTAGE STABILITY

Voltage stability is a slower phenomenon than transient stability, taking normally a few minutes to make the system unstable after contingencies. Despite the difference in timescale, operators do not have sufficient time to steer the system away manually from instability once the contingency occurs. For that reason, the process of posturing

Practical Power System Operation, First Edition. Ebrahim Vaahedi.
© 2014 The Institute of Electrical and Electronics Engineers, Inc. Published 2014 by John Wiley & Sons, Inc.

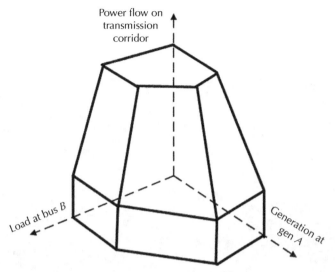

Figure 6.1   Acceptable operating conditions under normal conditions to avoid transient instability under contingencies.

the system for voltage stability involves developing preventive and corrective measures without any manual interaction from operators after the contingencies.

The preventive measures include creating constraints for the precontingency operating conditions called "nomograms" as shown in Figure 6.1. These nomograms restrict precontingency operating conditions to avoid transient instability after contingencies.

Utilities also deploy corrective actions to avoid voltage stability. These corrective actions must be designed and implemented under normal operating conditions to avoid voltage instability following contingencies. Through experience and knowledge about their systems, utilities develop remedial action schemes (RAS) for contingencies that normally cause operational infeasibility. These RAS normally involve automatic initiation of system or network changes that need to take place after sensing a contingency. For example, an RAS could involve switching in a number of capacitor banks and switching out a number of reactor banks. These remedial action schemes should be in place to provide comfort to the operator to ensure that the system is OK after any credible contingencies.

In summary, operators are provided with a combination of precontingency nomograms or operating condition constraints as well as corrective actions that must be activated to ensure system transient stability. Operators will need to ensure that the prevailing operating condition remains within the restricted nomograms or constraints. As well, operators may be provided with the corrective actions or RAS under different contingencies that they need to approve for activation under different contingencies. Figure 6.2 depicts one of the voltage stability displays available to operators at BC Hydro. This display provides the operators with one nomogram restricting precontingency operating conditions to ensure system voltage stability under different contingencies. This nomogram is obtained with respect to two major generation plants at BC Hydro.

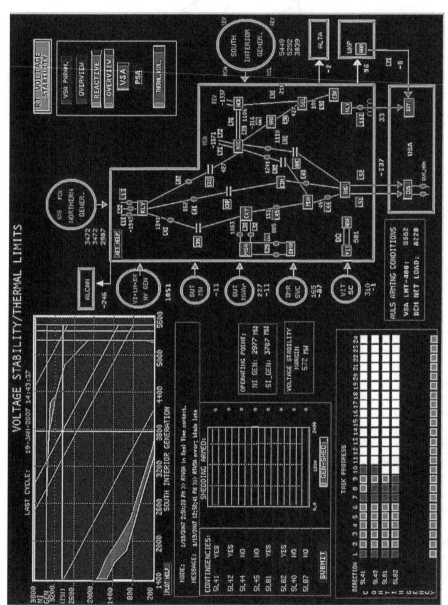

Figure 6.2 A typical EMS offline-based voltage stability display.

6.3 TECHNOLOGY FOR POWER SYSTEM POSTURING: VOLTAGE STABILITY

There are two sets of technologies needed for voltage stability posturing:

1. Voltage stability assessment
2. Implementation of voltage stability limits and remedial action schemes

6.3.1 Voltage Stability Assessment

Voltage security deals with all aspects of maintaining acceptable voltages. This can be divided into two areas:

- *Maintaining acceptable voltage profiles*, which includes acceptable pre- and postcontingency voltages, acceptable postcontingency voltage change, and acceptable postcontingency transient voltage variations.
- *Maintaining voltage stability*

Voltage stability is concerned with the ability of power systems to maintain steady acceptable voltage at all buses under normal and contingency conditions [1]. A system enters a state of voltage instability when a disturbance, such as increase in load or change in system condition, causes progressive and uncontrollable decline in voltage. The main factor causing instability is the inability of power systems to meet demands for reactive power. There are a number of changes and control devices in power systems, which contribute to voltage stability.

- Generators, synchronous condensers, or SVCs reaching reactive power limits
- Action of tap-changing transformers
- Load recovery dynamics
- Line tripping or generator outages

Recognizing the effect of changes and controls on voltage stability, Reference 33 defines voltage stability as

voltage stability stems from the attempt of load dynamics to restore power consumptions beyond capability of the combined transmission and generation system.

To understand voltage stability, we start with two simple systems to establish their power load characteristics and deduce the large system characteristics and acceptable region of operation.

6.3.1.1 Simple Resistive System   Figure 6.3 shows a simple resistive system supplying a load of R_{load}
We derive the relationship between P and V_{load} as

$$P = I_{load}V_{load} = \frac{\left(E - V_{load}\right)}{R}V_{load} = \frac{E}{R}V_{load} - \frac{1}{R}V_{load}^2$$

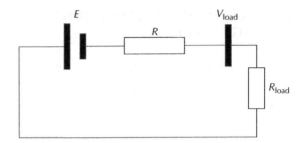

Figure 6.3 Simple resistive network.

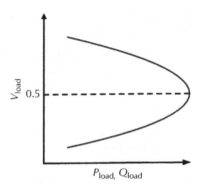

Figure 6.4 Resistive network voltage load characteristic.

Providing the power voltage characteristic given in Figure 6.4, it is clear that there are two voltages for every value of power; one on the top portion of the curve and the other one in the bottom. The maximum power P_{max} can be obtained by taking the derivative of P with respect to R_{load} as

$$P_{load} = I^2 R_{load} = \left[\frac{E}{\left(R + R_{load}\right)}\right]^2 R_{load} = E^2 \left[\frac{1}{\left(R + R_{load}\right)^2}\right] R_{load}$$

$$\frac{\partial P}{\partial R_{load}} = -2E^2 \left[\frac{1}{\left(R + R_{load}\right)^3}\right] R_{load} + E^2 \left[\frac{1}{\left(R + R_{load}\right)}\right]^2 = 0$$

from which we get

$$R_{load} = R$$

$$P_{max} = E^2 \left[\frac{1}{\left(R + R_{load}\right)^2}\right] R_{load} = E^2 \left[\frac{1}{\left(R + R_{load}\right)^2}\right] R_{load} = \frac{E^2}{4R}$$

6.3.1.2 Simple Two-Bus System

Consider the following two-bus system (Figure 6.5):

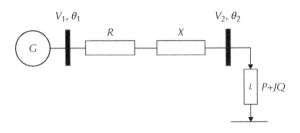

Figure 6.5   Simple two-bus system.

Using power flow Equations 3.33 and 3.34, we get

$$P_2 = G_{21}V_1V_2 \cos(\vartheta_2) + B_{21}V_1V_2 \sin(\vartheta_2) + G_{22}V_2^2 \tag{6.1}$$

$$Q_2 = G_{21}V_1V_2 \sin(\vartheta_2) - B_{21}V_1V_2 \cos(\vartheta_2) - B_{22}V_2^2 \tag{6.2}$$

As will be shown later in this section, the relationships for P_2 and Q_2 given in Equations 6.1 and 6.2 lend themselves to P–V and Q–V characteristics similar to the one given in Figure 6.1.

To create the two relationships P_2–V_2 or Q_2–V_2, we can solve the two Equations 6.1 and 6.2 for different values of V_2. Another method is to eliminate (ϑ_2) from the two equations.

To simplify the process of elimination, let us assume that the system is lossless ($R=0$). This assumption simplifies the two equations to

$$-P_2 = B_{21}V_1V_2 \sin(\vartheta_2) \tag{6.3}$$

$$-Q_2 = -B_{21}V_1V_2 \cos(\vartheta_2) - B_{22}V_2^2 \tag{6.4}$$

Knowing that

$$B_{21} = \frac{-1}{X_{21}} \quad \text{and} \quad B_{22} = \frac{-1}{X_{22}}$$

We get

$$-P_2 = \frac{V_1V_2}{X_{21}} \sin(\vartheta_2) \tag{6.5}$$

$$-Q_2 = \frac{V_1V_2}{X_{21}} \cos(\vartheta_2) - \frac{V_2^2}{X_{21}} \tag{6.6}$$

Note that the negative sign associated with P_2 and Q_2 is because the power flow equations assume injected P and Q as positive values. In this case, we assume P_2 and Q_2 to be consumed load extracted from the system.

By substituting $\sin(\vartheta_2)$ from the P_2 equation into the Q_2 equation, we get

$$-Q_2 = \frac{V_1V_2}{X_{21}} \sqrt{1 - \left(\frac{-X_{21}P_2}{V_1V_2}\right)^2} - \frac{V_2^2}{X_{21}} \tag{6.7}$$

This creates a second-order equation where the solution in terms of V_2 is

$$V_2 = \sqrt{\frac{\left(V_1^2 - 2QX\right) \pm \sqrt{V_1^4 - 4QXV_1^2 - 4P^2X^2}}{2}} \tag{6.8}$$

This relationship creates a characteristic similar to the one shown in Figure 6.2.

6.3.1.3 Voltage Stability Characteristics of a General System
These two examples attempt to indicate the general system behavior when loads are being supplied through transmission systems. These characteristics indicate that for every power level there are two voltage values, one of which is lower than the other one. Generally, systems are designed and operated so that they operate on the upper side of the curve for two reasons:

- The voltage on the lower part of the curve may not be in the acceptable voltage profile operating region for example, 0.9–1.1 p.u.

- The controls in power systems designed to restore the system inadvertently destabilize the system behavior.

We will explain the second point in Figure 6.6. Let us assume that the system is at operating point A. Now let us assume that the load increases and the load control mechanisms move the system to point B, where the voltage is less than A. This is the way that control devices such as ULTC are designed to operate. Once the load reduces, the control mechanisms on the system will take back the system load from B to A. Now let us assume that the system is at the operating point C. When the load increases, the load control mechanisms move the operating condition to point D, where both the load P and the voltage are actually lower instead of moving it to an operating point with larger P. As load power P becomes smaller, the load control

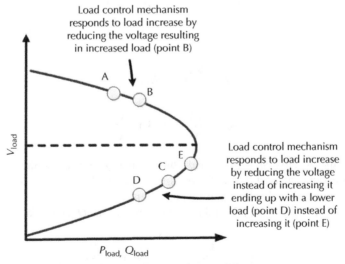

Figure 6.6 Impact of load increase on the operating condition.

mechanisms in the system continue to push the operating point even further away from point C to points with even lower *P* and *V*, making the system unstable [33].

The aforementioned discussion reinforces the point that the design characteristics of power system devices and controls force its operation to the region on the upper side of the *P–V* or *P–Q* curve.

6.3.1.4 Voltage Stability Criteria   Voltage stability for a general system involves obtaining the maximum limit or the nose point of the normal system (point D on Figure 6.7) and the nose points of all other critical contingencies and ensures that the existing operating condition is acceptable under normal and emergency conditions. Figure 6.5 elaborates on the necessary condition for voltage stability. Let us assume that the operating condition during normal condition is A. After the contingencies, this operating condition will move to points G or H. The normal system characteristic shows that we can stress the operating condition all the way to point D. However, the contingency with the smallest nose called the most critical contingency restricts the operating condition to point C because after this contingency the operating condition will jump to point F, which is still acceptable.

Guidelines have been developed in the past 2 decades [31–36] to guard against voltage stability using precontingency measures such as reactive reserve system wide or at a number of key generators, voltages, interface flows, or total load in an area. Using such criteria, utilities attempt to stay away from the nose of the precontingency curve (point D, Figure 6.5) by a certain margin (point C, Figure 6.5) to avoid voltage stability following the critical contingencies. The methods adopted will depend largely on the utilities' experience, policies, and regulatory requirements. For example, if studies show that voltage instability may occur when reactive reserves on specific generators reach certain values, the utility may use such measures as direct indicators of voltage security. The success of any such method depends on an understanding of the mechanism of, and proximity to, voltage instability for the particular

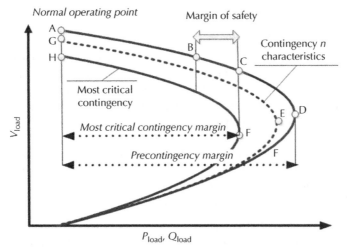

Figure 6.7   Voltage stability margin.

system under a wide variety of possible conditions. This section provides some generalized guidelines for developing and applying security assessment methods.

Some utilities add an extra safety margin to cater to the calculation data and modeling approximations. This safety margin moves the maximum operating condition from point C to point B in Figure 6.7.

Utilities may have different practices in calculating voltage stability. To obtain the curves, utilities can choose to increase different stress variables such as

- Only system-real load or P/V curves
- Only system-reactive load or Q/V curves
- Proportional increase in real and reactive load $P+jQ/V$

It can be stated that "the system must be operated such that, for the operating point and under all credible contingencies, the voltage stability margin remains larger than $x\%$ of the corresponding stress variable where $x\%$ is the safety margin."

For example, when the stress variable is defined as the area load and the criterion is defined as 7%, the system must remain voltage-stable under all contingencies when the area load is increased by 7% above the given operating level.

In addition to the criterion for voltage stability margin, utilities may establish other operating criteria for voltage security, such as

- Voltage decline/rise criteria, which specify that bus voltages must remain within $+x\%$ and $-y\%$ of the nominal (or precontingency) values under all contingencies
- Reactive reserve criteria, which specify that the reactive power reserve of individuals or groups of VAr sources (generators and controllable shunts) must remain above $x\%$ of their reactive power output under all contingencies.

6.3.1.5 Voltage Stability Assessment
In practice, there are two categories of voltage stability assessment methods:

- Power flow-based methods
- Time domain-based methods

Both of these methods attempt to establish whether there is an acceptable operating condition for an operating point to establish the load voltage characteristics. While the power flow-based methods establish the operating condition ignoring the transients from moving from one operating condition to the next, the time domain simulation tools simulate the dynamic performance of the system.

Power Flow-Based Methods   The power flow-based methods assume that the transients associated with system dynamics have died out, treating each new operating condition as a steady-state operating condition taking into effect the control actions. There are two methods under this category.

- *Power flow method*: The maximum loadability or the maximum value for the stress variable (P, Q, or $P+jQ$) can be calculated by starting at the initial operating point (point A in Figure 6.6) and making small increments in the stress variable

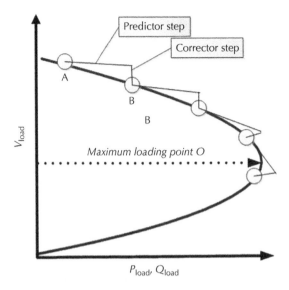

Figure 6.8   Continuation power flow method.

and make the corresponding increase in the generation pattern to reflect the appropriate conditions for scenarios at point B, C, and so on. This is followed up by solving the corresponding power flows for each scenario. This will eventually lead to the maximum loading point O, beyond which the power flow will diverge.

- *Continuation power flow method*: The continuation power flow [37] finds a continuum of power flow solutions using a given load generation direction change. The continuation power flow calculates the solution of a series of power flow equations reformulated to include a parameter that provides the degree that the continuation variable (e.g., load) changes. This method, which uses a predictive corrective scheme, is capable of finding the system voltage stability curve without becoming divergent because it can avoid the maximum loading point as shown in Figure 6.8. By avoiding divergence at the nose point, this method can provide a better prediction of the nose point.

Time Domain-Based Methods   Time domain-based methods simulate the system performance behavior taking into account the long dynamic response times associated with device control actions. While power flow methods calculate a power flow corresponding to the steady-state conditions at each operating condition (points A, B, and O in Figure 6.6), the time domain-based methods run a detailed simulation at each operating condition, taking into account the switching and control actions as shown in Figure 6.9 to indicate whether that operating condition is voltage-stable. The advantages of these methods are the modeling accuracy and the control and protection system simulation, which depend on the evolution of system operating condition. Furthermore, these methods provide the ability to interpret the results and see details of the evolving conditions.

As an example, let us consider a case with the objective of assessing whether a switchable capacitor bank can resolve a voltage stability problem. While the power

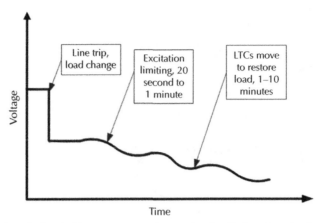

Figure 6.9   A typical switching sequence and control action in time domain simulation.

flow-based methods do not take into account the time delay associated with the operation of capacitor switching, the time domain-based methods simulate the operation of the capacitor switching to assess whether they operate in time to avoid voltage stability. Also, the study will provide more information on the interaction of switchable capacitor operation with other devices such as ULTCs.

There are two methods under this category.

- **Long-term time domain simulation**: In this method, the dynamic and static equations are fully solved to obtain the system performance [31] taking into account all the fast-response and slow-response devices participating in the dynamics of voltage stability. Since the simulations involve the actions of slow-response controllers such as transformer load tap changers (LTC), which operate to restore loads, the simulation time should span over a number of minutes. Performing time domain simulation for 10 minutes requires significant computation time. To reduce the computation time, variable time step methods have been suggested, which work well with implicit integration methods [31,33] creating very robust solutions. In these methods, by using a larger time step and a robust integration method, the short-term transients are filtered out.

- **Quasi-dynamic simulation**: In this method [38], the short-term transients are removed by substituting the dynamic response equations by their steady-state equilibrium equations. In effect, the differential equations representing the short-term transients are replaced by its steady-state algebraic equations. This method removes the short-term dynamics of transients while taking into account the dynamics of slow-response equipment such as LTCs. The time step for this method can be in the range of seconds compared to milliseconds for accurate time domain simulations, reducing the computation requirement by about 50–100 times providing a compromise between computational efficiency and accuracy. The disadvantage of this method is in its assumption that the short-term transients can be neglected with no ability to predict short-term transient voltage stability.

6.4   VOLTAGE STABILITY LIMIT DERIVATION AND IMPLEMENTATION

Due to computational burden, utilities precalculate voltage stability limits for different periods of the year and implement those in EMS to monitor and take the necessary action to mitigate against instability. In recent years though, online voltage stability has been implemented to make the calculations in real time with more accuracy. This new development will be covered in Chapter 9.

6.4.1   Voltage Stability Limit Derivation

Similar to transient stability limit derivation, the idea is to select a number of system stability features and translate the system voltage stability to those features. These system stability features are normally selected from prominent system variables that play a major role in system voltage stability such as

- The total power flow on a branch, an interface, or a corridor
- The total generation at a plant
- The total load in one area
- System voltage at a specific system location

To formulate operating guidelines that ensure the secure operation of the system, a large number of offline voltage stability studies are conducted to ensure that we find the operating parameter ranges for which the normal system and the system after all critical contingencies remains voltage-stable. Figure 6.10 shows the acceptable region of voltage stability shown for the two operating parameters 1 and 2. This area represents the common acceptable area between all of the voltage stability contours of the normal system and those of the contingencies. A number of these nomograms

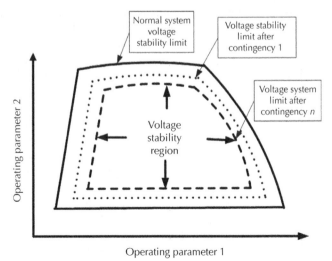

Figure 6.10   Acceptable region of voltage stability in terms of the two operating parameters.

are provided to the operator to ensure that the system is operated in the acceptable area shown.

The following steps provide a typical sequence of offline-based voltage studies to provide acceptable regions of operation for a study period.

1. Choose the worst operating condition for the period to study. This includes the primary demand, generation pattern, and transactions with adjacent utilities. For example, if the studies are being done for the winter season, the worst operating condition for this period is selected.

2. Define the critical contingencies for which voltage stability assessments are needed. Technically, it would be better to use a contingency screening and ranking method to establish a short list of critical contingencies [39].

3. Establish the voltage stability of the system normal and the system after all the n critical contingencies at the present operating condition. If the system is voltage stability-unstable at the initial operating condition, the mitigating control action needs to be established (e.g., load shedding) to make the system stable.

4. Determine the acceptable operating limit by increasing the operating parameters until the system becomes unstable following one of the critical likely contingencies.

In this list, item 3 should be repeated until two cases are obtained, one unstable and one stable. The stable case represents a case before reaching the nose point whereas the unstable point represents a point that is beyond the nose point. The value of the operating parameter is determined, and the limit will be the value determined minus a confidence margin established by the utility.

6.4.2   Implementation of Voltage Stability Limits

Once the system operating limits are determined from offline studies, these limits are then formulated in the form of nomograms in the energy management system. Figure 6.11 shows one of the voltage stability displays available to operators at BC Hydro. This display provides the operators with two nomograms for system normal conditions restricting precontingency operating conditions to ensure system voltage stability under different contingencies. The nomogram on top provides the region in terms of NI generation versus SI generation while the nomogram in the bottom provides the region in terms of net system load versus SI generation. Figure 6.12 provides the same nomograms for the case when one major transmission is placed out of service.

The energy management system calculates from prevailing system conditions the values of the operating parameters and compares them against the limits. It warns the operator in case an operating parameter is likely to be violated. If the existing system is found to be outside the stable operating region, preformulated remedial actions such as load shedding can be armed in the SCADA. For example, the prevailing system condition is shown in Figure 6.11, indicating that it is well within the acceptable region of operation. On the other hand, the operating condition in

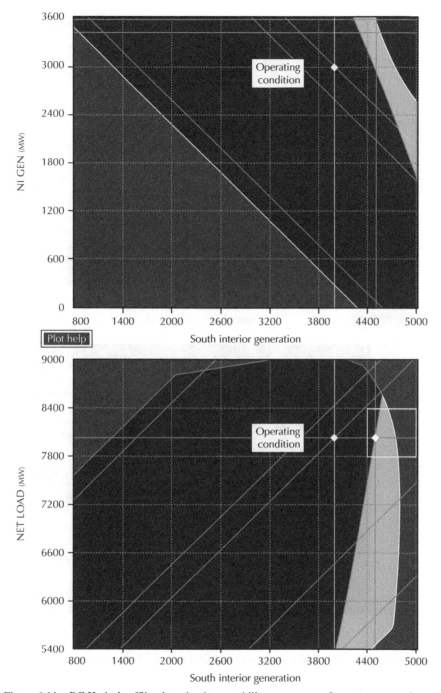

Figure 6.11   BC Hydro's offline-based voltage stability nomograms for system normal.

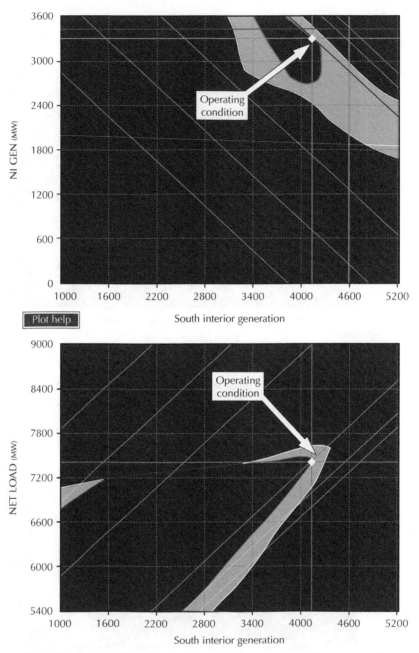

Figure 6.12   BC Hydro's offline-based voltage stability nomograms for system with one element out of service.

Figure 6.12 is on the border of voltage stability region. If the prevailing operating condition moves outside the region, the application will send a warning to the operator. Furthermore, the system may take automatic remedial action if such actions have already been preformulated in the SCADA.

QUESTIONS AND PROBLEMS

6.1. Define power system voltage stability.

6.2. List the two conditions required for power system security.

6.3. List and describe two sets of technology needed for voltage stability posturing.

6.4. Describe the time domain-based methods for power system voltage stability assessment.

6.5. List and describe the steps for a typical sequence of offline-based voltage studies to provide acceptable regions of operation for a study period.

6.6. What is stress parameter?

6.7. Label the following graph with the following (Figure 6.13):

- Normal operating point
- Margin safety
- Most critical contingency
- Precontingency margin

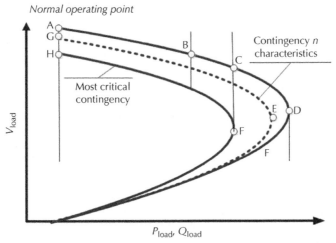

Figure 6.13 Precontingency and postcontingency voltage stability characteristics for Question 6.7.

6.8. Using the following analytical equation for the two-bus system given in Figure 6.5, derive the nose point assuming $Q=0.3$, $X=0.1$ and $V_1=1.06$.

$$V_2 = \sqrt{\frac{\left(V_1^2 - 2QX\right) \pm \sqrt{V_1^4 - 4QXV_1^2 - 4P^2X^2}}{2}}$$

6.9. Using the equation given in Question 6.8, plot the V_2–P_2 curves for the following cases:

Case	V_1	Q	X
1	1.00	0.436	0.4751
2	1.00	0.436	0.46
3	1.00	0.436	0.45

6.10. A utility has a normal operating condition of $P_{load} = 1000$ MW with the following voltage stability limit values under four critical contingencies.

a. How much load can the system supply if no contingencies are considered?

b. How much voltage stability margin does the system have if no contingency case is considered?

c. How much load can the system supply if the system voltage stability limits under contingencies are also considered?

d. How much load would the system supply if the operator uses a safety margin of 80 MW?

System condition	Voltage stability limit
Normal system	2000 MW
Contingency 1	1800 MW
Contingency 2	1400 MW
Contingency 3	1100 MW
Contingency 4	1300 MW

6.11. A utility has a normal operating condition of $P_{load} = 1000$ MW, 500 MVar with the following voltage stability limit values under four critical contingencies:

a. What is the voltage stability margin if no contingency cases are considered in terms of different stress parameters?

b. What is the voltage stability margin if the system voltage stability limits under contingencies are also considered in terms of different stress parameters?

System condition	Voltage stability limit	Voltage stability limit	Voltage stability limit
Stress parameter	P	Q	P, Q
Normal system	2000 MW	1500 MVar	2000 MW, 1500 MVar
Contingency 1	1800 MW	1300 MVar	1800 MW, 1300 MVar
Contingency 2	1400 MW	1000 MW	1400 MW, 1000 MVar
Contingency 3	1100 MW	800 MW	1100 MW, 800 MVar
Contingency 4	1300 MW	1100 MVar	1300 MW, 1100 MVar

CHAPTER 7

POWER SYSTEM GENERATION LOAD BALANCE

7.1 OPERATOR'S QUESTION ON GENERATION LOAD BALANCE

A fundamental responsibility of system operators is to balance load and generation in real time, where load represents the summation of system native load and the scheduled exchanges to other utilities. The North American Reliability Council (NERC) has come up with standards to ensure that each utility fulfills its load and generation balance so that it does not place undue balancing responsibility on other utilities.

To balance load and generation in real time, the system operator needs to ensure that the automatic generation control (AGC) system that regulates a number of generating units to match generation to load is working within NERC's established reliability standards.

7.2 PROCESS FOR GENERATION LOAD BALANCE

7.2.1 Introduction

Utilities need to plan and operate their system to ensure that they can supply their loads. There are four distinct time frames in ensuring that they can supply their loads as shown in Figure 7.1. Long-term planning ensures that the most optimal generation portfolio is invested in to supply the forecasted load. Operations planning deals with changes in transmission or generation that will need to take place for maintenance purposes in the coming months. The unit function that deals with the optimum selection of the units that need to go online to supply the load may fall in this time frame as well, depending on their type, generating units need different preparation time for going live spanning from months to days. For example, while nuclear units need months of preparation, hydraulic units can go live in a day or so. Economic dispatch deals with the selection of the most economic units to supply the load in the next few hours.

The objective of AGC, which is also called as load frequency control, is to balance generation and load on a minute-to-minute basis when operators do not have

Practical Power System Operation, First Edition. Ebrahim Vaahedi.
© 2014 The Institute of Electrical and Electronics Engineers, Inc. Published 2014 by John Wiley & Sons, Inc.

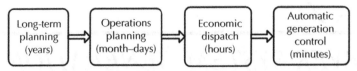

Figure 7.1   Generation load balance in different time horizons.

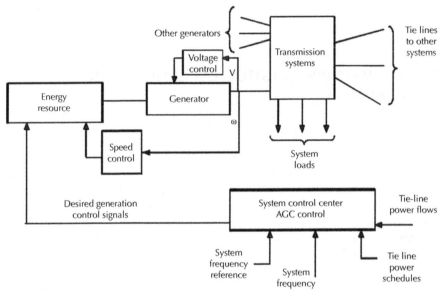

Figure 7.2   Power system automatic generation control.

sufficient time to control generators. This objective is derived from the following tenets of an interconnected power system operation:

1. Each system should provide sufficient capacity to supply its load under normal conditions.

2. Each system should have sufficient capacity to supply its scheduled interchange obligations to other interconnected utilities.

3. Each system should provide its own regulation service requirement not imposing any regulation burden on other systems.

4. Each system should synchronize its time to exact time, getting rid of the accumulated frequency deviations over time.

5. Each system should provide sufficient capacity during contingencies to balance its load. However, during the contingency conditions, the system can rely on its interconnection to provide some capacity support.

To achieve these objectives, a new centralized controller called AGC, as shown in Figure 7.2, is developed, which works along other generator voltage and speed control systems. As Figure 7.2 shows, AGC measures actual system frequency and interchange flows from which it calculates the frequency and interchange flow deviations by using the reference frequency and scheduled interchange values. The frequency and interchange deviation are then used to balance load and generation on a minute-to-minute basis.

Considering that each utility's load is composed of its native load and the scheduled transactions with other utilities, two system measurements are used to reflect the degree of generation load imbalance. The first indicator is the system frequency, which is constant in the whole system and reflects whether the native load is balanced. This can be best described by considering an isolated utility with no interties to other areas. By removing the frequency error, the AGC control ensures that the generation and load are balanced out.

The second indicator is the total interchange obligation the utility has to other utilities. So if a number of interconnected utilities all attempt to remove the system frequency deviations, they will end up balancing load and generation for the whole interconnected system, except that the final outcome may not fulfill their interchange obligations to each other. For that reason, the AGC controller must consider the minimization of both frequency and interchange deviations as its objective function in its control design. Hence, area control error (ACE) is defined for a system as

$$ACE = -10 \cdot \beta_f \cdot (f_{Actual} - f_{Desired}) + (T_{Actual} - T_{Scheduled}) \qquad (7.1)$$

where, $f_{Desired}$ is the desired frequency, Hz (e.g., 60 Hz); f_{Actual} is the actual frequency, Hz; T_{Actual} is the actual interchange schedule or tieline flow with a positive sign for export, MW; $T_{Scheduled}$ is the scheduled interchange or tieline flow with a positive sign for export, MW; β_f is the area bias with a negative value, MW per 0.1 Hz

The first term of ACE shows how well the native load and generation balance each other out. If the actual frequency is larger than the desired value, this creates a positive term, indicating overgeneration and vice versa. Equally, the second term shows the deviation of the system's actual interchange flow from the desired value for each control area. If, for example, the actual exported tieline flow for one area is more than the scheduled amount, the ACE is again positive, indicating overgeneration in that area.

AGC achieves its objective by minimizing or bounding ACE. It is important to note that ACE only considers the status of a snapshot of the system disregarding the frequency deviations in the previous time periods and the accumulated frequency and interchange flow mismatches. This creates other issues associated with accumulated frequency and interchange deviations as shown in Figure 7.3.

As shown in Figure 7.3, while frequency deviation and interchange deviations have been reduced to zero by AGC control, there are accumulated frequency and interchange deviations. All electric clocks fed by the main system are based on system frequency. The time error is a function of accumulated frequency deviation reflected by the area under the frequency deviation curve in Figure 7.3. To ensure that the time shown by electric clocks is correct, system operators should correct the

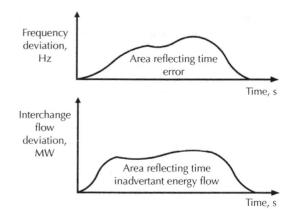

Figure 7.3 Frequency and interchange flow deviations for one area.

accumulated time error by scheduled actions. The time error resulting from frequency deviation can be calculated by

$$\text{Time error} = \int \left(\frac{f_{\text{Actual}} - f_{\text{Desired}}}{f_{\text{Desired}}} \right) \cdot dt \qquad (7.2)$$

where f_{Actual} and f_{Desired} are in Hz.

While the time error adjustment to zero is not accomplished by AGC, it is sometimes added to the ACE Equation 7.1 for the sake of completeness, as given in Equation 7.3:

$$\text{ACE} = -10 \cdot \beta_{\text{f}} \cdot (f_{\text{Actual}} - f_{\text{Desired}}) + (T_{\text{Actual}} - T_{\text{Scheduled}}) + \beta_{\text{t}} \cdot (\text{Time error}) \qquad (7.3)$$

where β_{t} is the time error bias in MW per second.

The accumulated deviation between the scheduled and actual intertie flow is called "inadvertent flow," which needs to be dealt with financially by the utilities involved. For example, the accumulated interchange deviation in Figure 7.3 reflected by the area under the interchange curve shows the additional energy supplied by area 1. There is also a process between interconnected utilities to deal with the inadvertent flow on a periodic basis.

7.2.2 NERC Standards for Automatic Generation Control

NERC has established two standards, CPS1 and CPS2, for AGC operation in the interconnected North American system [40]. In simple terms, CPS1 reflects average of the 1 minute frequency changes in one balancing area divided by the same value averaged over the whole interconnected system consisting of different balancing areas. Balancing area refers to an individual area that is responsible for its load generation balance. CPS2 is similar to CPS1 except that the averages are conducted for six 10 minute periods in every hour.

Both these standards have a minimum value that utilities would need to achieve for their systems. In other words, the AGC standards are put in place to ensure that each utility provides its fair share of regulation.

NERC requires each utility to have reliable technology to measure ACE and implement AGC functions. Furthermore, NERC requires utilities to retain information on AGC for 1 year and report the AGC performance information as needed. Furthermore, NERC guidelines [40] also define 10 minutes as the maximum duration for an area to return its ACE to zero from the beginning of a disturbance in that area. They also encourage each area to control its ACE to cross zero at least every 10 minutes and have a 10 minute average below the average of specific interconnected system comprising a number of areas.

In 2010, NERC started exploring a more accurate standard for assessing AGC performance called BAAL [41]. While as of the date of writing this chapter BAAL has not been approved yet, it is expected that in the future more accurate standards will replace the existing standards.

7.2.3   Process for Automatic Generation Control

Operators need to monitor system frequency, tieline flows, scheduled tieline flows, and ACE on a minute-to-minute basis as shown in Figure 7.4. They need to examine the frequency deviations, the tieline flow deviations from the scheduled flows, and ACE to ensure that they all stay within acceptable thresholds and that the AGC is operating properly. If the operator finds out that system frequency or ACE has violated acceptable thresholds, then they need to intervene and change system generation to bring back frequency deviation and ACE to the acceptable range. For example, Figure 7.4 shows BC Hydro's frequency variation, load variation, and actual interchange flows and their corresponding interchange schedules. Another duty of operators is to ensure that the AGC performance information, including the values for CPS1 and CPS2, is retained up to a year for future auditing and examination by NERC. Figure 7.5 shows another AGC display providing more details on AGC generation fleet and generation reserve

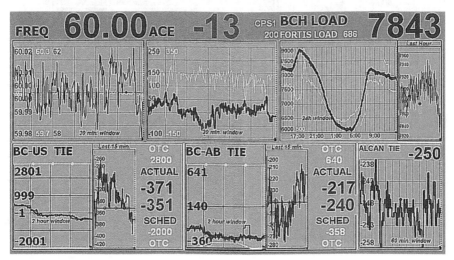

Figure 7.4   An operator AGC display.

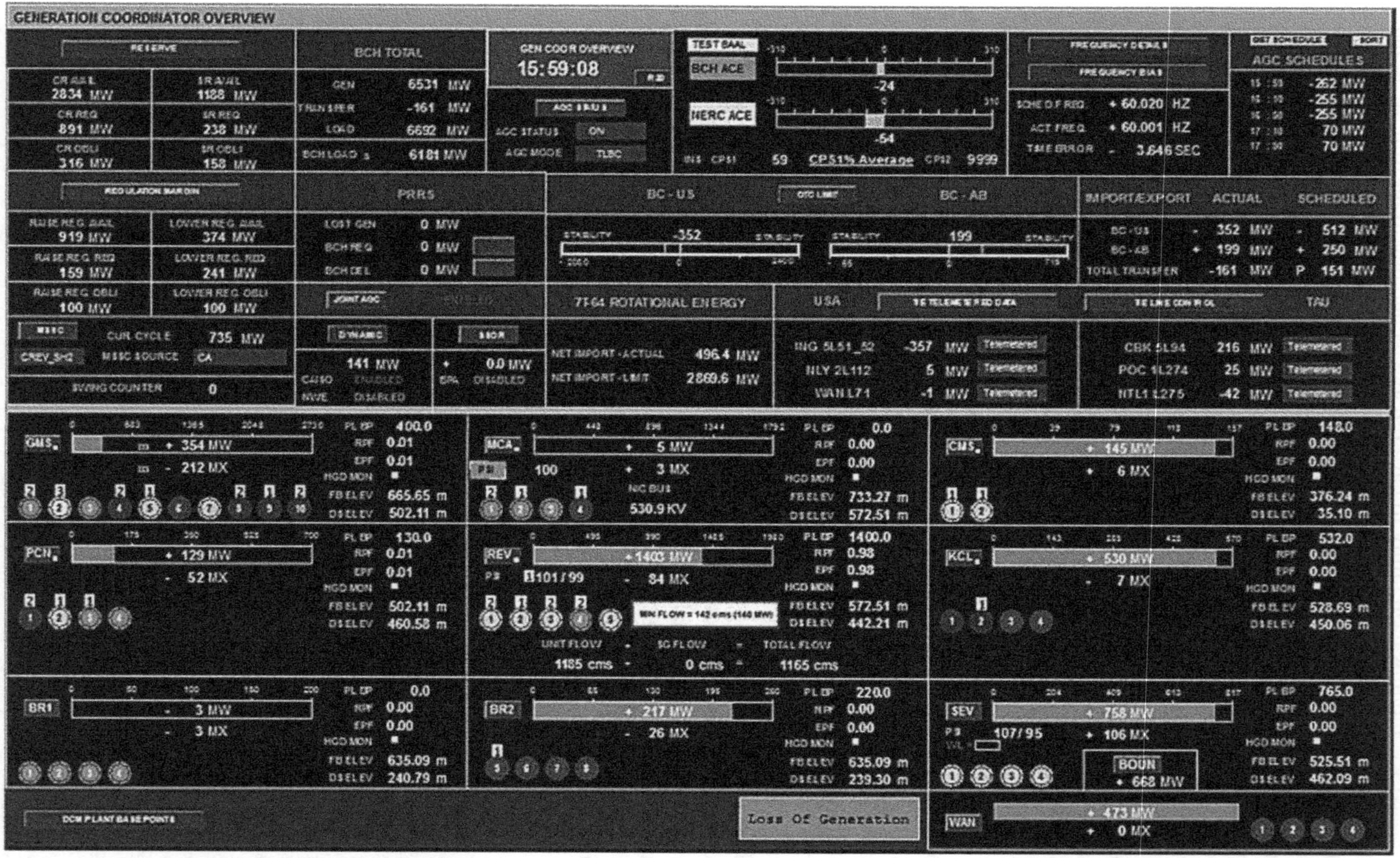

Figure 7.5 BC Hydro's AGC display with AGC generation fleet and AGC reserve requirements.

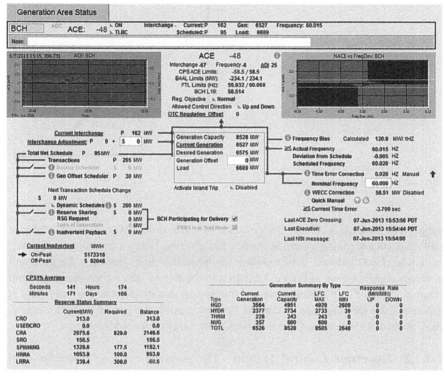

Figure 7.6   BC Hydro's AGC display for BAAL standard.

requirements. Figure 7.6 shows a more recent AGC display reflecting the new BAAL standard for which is aimed to better reflect AGC performance.

7.3   TECHNOLOGY FOR GENERATION LOAD BALANCE

NERC requires each system or Balancing Authority [40] to have a very reliable AGC system with a high availability while AGC availability means the percentage of time that the system is in service (e.g., minimum 99.95% availability) with ACE measurements provided at least once every few seconds (e.g., 6 seconds). AGC system consists of two technology elements:

1. AGC application
2. AGC infrastructure

7.3.1   Automatic Generation Control Application

7.3.1.1   Introduction   To understand the operation of AGC [42–45], one needs to understand the natural system frequency response and the additional control needed to capture frequency change as system condition changes. When load in a system increases,

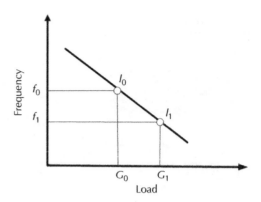

Figure 7.7   A typical speed power
characteristic of a governor system.

the power mismatch is initially compensated by using system kinetic energy, which in turn results in a declining system frequency. Since system load is a function of frequency, system load gets reduced, compensating the frequency drop. However, the generator governor control shown in Figure 7.2 comes into action to increase generation to limit frequency drop. The governor control action results in a steady-state frequency that is different from the reference or desired frequency. After the governor control action, AGC is needed to bring the frequency back to the original desired frequency.

7.3.1.2   *Governor Control System*
The governor control system for each generator, as shown in Figure 7.2, measures system frequency and feeds back to the generator prime-mover control the frequency deviation changing the input mechanical power to the generator. So if the frequency is lower than the synchronous frequency f_0, the deviation results in an increase in the generator mechanical power. A typical speed power characteristic of a governor system is shown in Figure 7.7.

This figure shows the generation response to the system frequency change as a result of governor action. For example, at point I_0, the frequency deviation is zero, corresponding to the balance between generation and load with zero adjustment in generation change. On the other hand, at point I_1, the frequency deviation of $(f_0 - f_1)$ results in a generation increase of $(G_1 - G_0)$. In other words, if there is an imbalance of $(G_1 - G_0)$ between generation and load, the system frequency will drop to f_1, resulting in the governor action to remove the imbalance by increasing the generation to G_1.

The slope of this curve is called governor droop.

$$\text{Governor Droop} = -\frac{\Delta f}{\Delta G} \text{ in Hz/MW} \tag{7.4}$$

$$\text{p.u. Governor Droop\%} = -\frac{\Delta f / f_0}{\Delta G / G_R} \tag{7.5}$$

where f_0 is the rated frequency in Hz and G_R is the rated generation capacity in MW.

The higher the droop, the less sensitive the generation response is to a frequency change. Since the system frequency is constant all over the system, this relationship shows that for a change in system frequency Δf, different generating units respond to the frequency change inversely proportional to their governor droops:

$$\frac{\Delta G_i}{G_R} = \frac{-(\Delta f / f_0)}{GD_i} \Rightarrow \Delta G_i = -(\Delta f)\frac{(G_R / f_0)}{GD_i} \qquad (7.6)$$

where ΔG_i and GD_i are generation change and the governor droop of generator i respectively. The total generation change in an n generator system as a result of the frequency change will be

$$\text{Total generation change in MW} = \sum_{i=1}^{n} \Delta G_i = \sum_{i=1}^{n} -(\Delta f)\frac{(G_{Ri} / f_0)}{GD_i} \qquad (7.7)$$

It must be noted that as a result of the frequency change the system load also changes due to load sensitivity to frequency. The total system load change (ΔL) as a result of the change in frequency (Δf) can be obtained as

$$\Delta L = D \cdot \Delta f \qquad (7.8)$$

where ΔL_i and Δf are load and frequency changes in MW and Hz respectively and D is the load frequency variation factor in MW/Hz. Combining Equations 7.7 and 7.8, the system load generation imbalance equation can be derived as

$$\text{Total generation load change} = \sum_{i=1}^{n} \Delta G_i - \Delta L_i = \sum_{i=1}^{n} -(\Delta f)\frac{(G_{Ri} / f_0)}{GD_i} - D \cdot \Delta f \qquad (7.9)$$

or

$$\text{Total generation load change} = \sum_{i=1}^{n} \Delta G_i - \Delta L_i = -(\Delta f)\left[\sum_{i=1}^{n} \frac{(G_{Ri} / f_0)}{GD_i} + D\right] \qquad (7.10)$$

In other words, as a result of a frequency change (Δf), there is an adjustment in both load and generation given by Equation 7.10.

For example, if there are three units with p.u. droops of 1, 2, and 5% with the ratings of 100, 200, and 500 MW respectively and load variation factor of $D = 200$ MW/Hz, the total generation load change after a reduction of 0.1 Hz system frequency will be

$$\text{Total generation load response} = \left[-(-0.1)(100/60)/0.01\right] + \left[-(-0.1)(200/60)/0.02\right]$$
$$+ \left[-(-0.1)(500/60)/0.05\right] - (-0.1)(200)$$
$$= 50 + 20 = 70 \text{ MW}$$

This example demonstrates that due to the frequency change of 0.1 Hz the system generation increases by 50 MW and load gets reduced by 20 MW to create a total generation load change of 70 MW. To say it in a different way, if there is a frequency drop of 0.1 Hz in the system, the governor action increases generation by 50 MW to compensate for system generation deficit and the load will naturally reduce by another 20 MW.

7.3.1.3 Interconnected Operations Power systems interconnections are put in place for different systems to be able to

- Perform exchange of electricity and enjoy the economic benefits of diversity in generation and load
- Provide support under contingencies

Since load and generation in each system change instantaneously, it is important to have proper controls on interties. These controls ensure that the undesirable tieline flows do not show up as the systems try to mitigate frequency deviations. In other words, each system provides its share of frequency correction without impacting another system's generation load balance inadvertently [43,44].

7.3.1.4 AGC Application Automatic generation software is designed to provide supplemental signals to a number of generators called AGC units to minimize ACE. The supplementary signals are provided to the governor system as shown in Figure 7.2 of the regulating units to restore the frequency and substitute generation increase from governor-controlled units. The AGC software in effect uses system measurements such as system frequency, scheduled intertie flows, and actual intertie flows to build the value of ACE. Then, it uses this value to change the AGC unit generations proportional to their economic efficiency.

There are three different modes of operations for AGC:

1. Flat frequency
2. Flat tieline schedule
3. Tieline with frequency bias

The flat frequency mode is used by isolated systems that only need to worry about the system frequency, thus removing the term associated with the transaction errors in the ACE. If there are a number of interconnected areas with different AGC flat frequency control, then depending on the responsiveness of each area's AGC, the final share of generation to control a frequency change will be different. Let us assume that there are a number of areas with AGC flat frequency biases $\beta_f i$, then the contribution from each area is proportional to their $\beta_f i$, indicating the responsiveness of the area i in correcting the frequency. As an example, let us assume that there are three areas operating under flat frequency mode with contribution factors of $\beta_{f1} = 10$, $\beta_{f2} = 20$, and $\beta_{f3} = 30$. If we assume that there is a frequency drop of Δf in the system, then the ACE values in different areas become

Area 1 ACE$= -10\Delta f$
Area 2 ACE$= -20\Delta f$
Area 3 ACE$= 30\Delta f$

These values indicate that area 3 AGC generation response will be three times area 1's generation response while area 2's generation response will be twice that of area 1.

The flat tieline schedule mode ensures that the desired schedules are achieved, thus removing the frequency term from the ACE. In this mode, the system frequency deviation will not be corrected. Let us consider a case of two

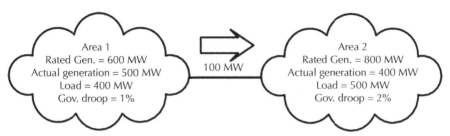

Figure 7.8   Two-area system.

areas system of Figure 7.8. Let us assume that there is 200 MW increase in load in area 1. Since there is no frequency control bias in ACE, areas 1 and 2 only attempt to retain the interchange schedule at 100 MW while not making any attempts to correct the frequency deviation resulting from the change in load. So in this case there will be a natural frequency drop in the system as a result of governor action resulting in

$$\text{Total generation load change} = 200 = -(\Delta f)\left[\sum_{i=1}^{n} \frac{(G_{Ri}/f_0)}{GD_i}\right]$$

$$200 = -(\Delta f)\left[\frac{(600/60)}{0.01} + \frac{(800/60)}{0.02}\right]$$

$$\Delta f = 0.12\,\text{Hz}$$

$$\text{Increased generation in area } 1 = (0.12)(600/60)/0.01 = 120\,\text{MW}$$

$$\text{Increased generation in area } 2 = (0.12)(800/60)/0.02 = 80\,\text{MW}$$

After governor operation with this generation increases, the interchange flow will have to change to 80 MW − 100 MW = −20 MW. In other words, because area 2 is generating 80 MW for the load increase in area 1, then instead of importing 100 MW, it will be importing 20 MW to area 1. After AGC operation, both areas try to reverse the interchange flow back to 100 MW export from area 1 to area 2 by changing their AGC unit generations. In other words, area 1 will increase its generation while area 2 will reduce its generation until the interchange mismatch is resolved. The frequency deviation of 0.12 Hz will not be corrected.

Finally, the tieline with frequency bias mode ensures that both the desired frequency and tieline schedule controls are achieved.

7.3.2   Automatic Generation Control Infrastructure

The software for AGC resides on the energy management system. The SCADA provides the infrastructure to measure the signals. EMS then forms ACE and provides the control signals for the AGC units using economic factors for different participating units. SCADA then is used to deploy the controls on the AGC units.

7.3.3 Example on AGC Operation

Example 7.1

In the system shown in Figure 7.9, assume that there is a 200 MW load increase in area 1. What would be the situation after the operation of (a) governor control and (b) AGC control with tieline and frequency bias, and (c) AGC control with flat frequency?

Solution

    a. After the operation of governor control, the loss of 200 MW should be supplied by increase in generation resulting from units. Using Equation 7.10, we get

$$200 = \sum_{i=1}^{n} \Delta G_i = \sum_{i=1}^{n} -(\Delta f)\frac{(G_{Ri}/f_0)}{GD_i}$$

$$200 = \sum_{i=1}^{3} \Delta G_i = -\Delta f\left(\frac{(G_{Ri}/f_0)}{GD_1} + \frac{(G_{R2}/f_0)}{GD_2} + \frac{(G_{R3}/f_0)}{GD_3}\right)$$

$$200 = -\Delta f\left(\frac{(1000/60)}{0.005} + \frac{(500/60)}{0.01} + \frac{(400/60)}{0.02}\right)$$

$$\Delta f = \frac{-200}{\left[(1000/60)/0.005 + (500/60)/0.01 + (400/60)/0.02\right]}$$

$\Delta f = -0.04444$ Hz

Area 1 Generation increase $= -(0.04444)(1000/60)/0.005 = 148.15$ MW

Area 2 Generation increase $= -(0.04444)(500/60)/0.01 = 37.04$ MW

Area 3 Generation increase $= -(0.04444)(400/60)/0.02 = 14.81$ MW

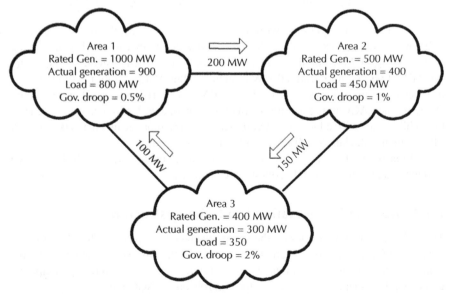

Figure 7.9 AGC example of three-area system.

b. After the operation of AGC system, the generation in area 1 will increase by 200 MW, removing the frequency deviation and keeping the interchange exchanges with areas 2 and 3 at the original values. There will be no change in the generation and load in areas 2 and 3.

Example 7.2

Let us assume that an area has three generators under AGC control with economic factors of 0.1, 0.3, and 0.6. These economic factors are used to ensure that the AGC generation dispatch among different units in one area is accomplished in an economic manner. Additional information about the system is given next:

Actual intertie flow = 1000 MW
Scheduled intertie flow = 1200 MW
Actual frequency = 59.5 Hz
$\beta_f = -100$ in MW per 0.1 Hz
What is the ACE and how much is generation increase at each unit?

Solution

Using Equation 7.1, we get

$$ACE = -10 \cdot (-100) \cdot (59.5 - 60) + (1000 - 1200)$$

$$ACE = -500 + (-200) = -700$$

So the total generation increase needed is 700 MW.

$$\text{Generator 1 increase} = 700 \cdot (0.1)/(0.1 + 0.3 + 0.6) = 70 \text{ MW}$$

$$\text{Generator 2 increase} = 700 \cdot (0.3)/(0.1 + 0.3 + 0.6) = 210 \text{ MW}$$

$$\text{Generator 3 increase} = 700 \cdot (0.6)/(0.1 + 0.3 + 0.6) = 420 \text{ MW}$$

QUESTIONS AND PROBLEMS

7.1. List four distinct time frames in the generation load balance time horizons.

7.2. What is AGC and what is its objective?

7.3. List five system operation tenets AGC is based on.

7.4. List three different modes of operations for AGC.

7.5. If there are four units with p.u. droops of 1, 23, and 4% with ratings of 200, 300, 600, 800 MW, and the total load frequency variation factor D of 200 Hz/MW, then what is the total generation change with 0.2 Hz system frequency change?

7.6. Assume there is a 200 MW load increase in area 1 of Figure 7.10. What would be the situation be after the operation?

a. Governor control

b. AGC control with interchange and frequency bias

c. AGC control with flat frequency bias

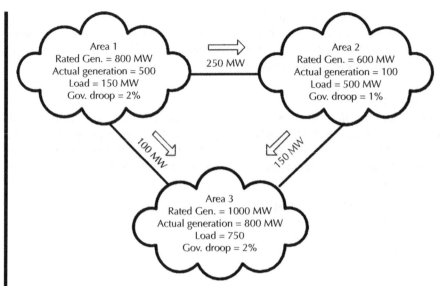

Figure 7.10 Three-area system.

7.7. Assume there is a 100 MW load increase in area 2 of Figure 7.11. What would be the situation be after the operation?

a. Governor control

b. AGC control with frequency and interchange bias

c. AGC with frequency control only in area 3

d. Assume that the frequency bias factors of $\beta_{f1}=10$, $\beta_{f2}=20$, $\beta_{f3}=30$, and $\beta_{f4}=40$ for different areas. With AGC control with frequency control in all areas, what would be the generation increase in different areas?

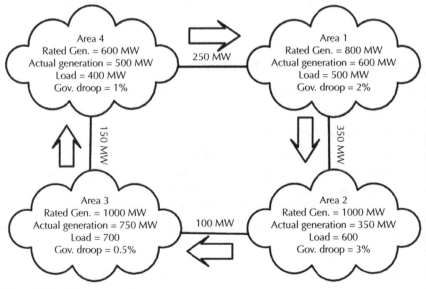

Figure 7.11 Four-area system.

CHAPTER *8*

POWER SYSTEM OPERATION OPTIMIZATION

8.1 OPERATOR'S QUESTION ON POWER SYSTEM OPERATION OPTIMIZATION

The main objectives of power system operation are

- Safety
- Reliability
- Efficiency

While safety and reliability are primary objectives, efficiency is also important. Efficiency means using process optimization while complying with reliability and safety requirements.

Similar to other objectives, efficiency applies to all segments of power system operation such as transmission system operation, distribution system operation, and generation operation.

The main application of optimization in generation operation is generation load sufficiency. Operators need to ensure that they have sufficient resources to operate the system in a reliable and economic manner. Within this context, the questions that operators may ask can be listed as

- Can a generating plant be selected for maintenance?
- What preparation should go on to bring plants online in the next few months?
- What units should remain in service for reliability reasons?
- How much reserve should the system have?
- Closer to real-time operation, what plants or generating units should be brought online and how much should each unit generate?
- Is the load and generation balanced in real time on a minute-to-minute basis?

Other applications of optimization in transmission and distribution operation include:

- Transmission loss minimization
- Voltage violations remedial action under normal and contingency conditions

Practical Power System Operation, First Edition. Ebrahim Vaahedi.
© 2014 The Institute of Electrical and Electronics Engineers, Inc. Published 2014 by John Wiley & Sons, Inc.

- Distribution voltage var optimization for energy conservation
- Distribution network reconfiguration to resolve voltage violations
- Transmission and distribution restoration

Since the application of optimization to power system generation operation constitutes the most significant element of power system optimization, the balance of this chapter discusses the application of optimization to the generation operation. However, the concepts discussed in this chapter are general and applicable to other applications of optimization in transmission and distribution area.

8.2 PROCESS FOR POWER SYSTEM GENERATION OPERATION

8.2.1 Introduction

Utilities need to plan and operate their system to ensure that they can supply their loads. The discipline for ensuring load and generation balance consists of four distinct time frames as shown in Figure 8.1.

Long-term planning ensures that the most optimal generation portfolio is planned, invested in, and developed in time to be able to supply the future load. Operations planning deals with changes in transmission or generation which will need to take place for maintenance purposes in the coming months. Operations planning also includes unit commitment function, which deals with the optimum selection of the units that need to go online to supply the load in a matter of months or days. Depending on their type, generating plants need different preparation time for going live. While nuclear plants need months of preparation, hydraulic units can go live in a matter of a day or so. Near real-time operation deals with selecting the most economic generation portfolio to supply the load (unit commitment) and the optimum set point for the units (economic dispatch) to supply that load for actual operation within a day or a few hours. Finally, real-time operation deals with preparations going on for the actual operation in a matter of minutes.

This chapter covers the operator's questions and the processes associated with the time frames of operations planning and near real time. The process of generation load balance in the real-time time frame was covered in Chapter 7.

8.2.2 Utility Model

Up until two decades ago, utilities in the world had a vertically integrated model. In the past two decades, to enhance competition and efficiency, a number of utilities in the world have restructured their model unbundling their functions.

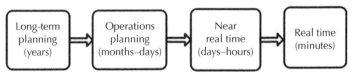

Figure 8.1 Generation sufficiency in different time horizons.

8.2.2.1 *Conventional Vertically Integrated Utility Model*   In this model, the utility performs all the functions of generation, transmission, and distribution of the electricity all the way to the end customers. The utility owns all the assets and performs operation and maintenance of the system. The utility would have to provide sufficient generation and transmission to supply the load in a safe, reliable, and efficient way.

8.2.2.2 *Restructured Utility Model*   Restructuring of the utilities started in the early 1990s to create competition and efficiency [46,47]. This resulted in unbundling of the vertically integrated utilities into many pieces, creating competition in generation and transmission systems. The four dominant features of the restructured systems are

1. **Independent system operator (ISO)** operates the system and provides non-discriminatory transmission services.
2. **Transmission market** administered by ISO for interchange energy scheduling external to the ISO region. The available intertie capacity is calculated and posted on a transmission reservation system called Open Access Same time Information System (OASIS). OASIS then can be accessed for transmission reservation by entities that want to move electricity from other regions through these interties.
3. **Energy market** provides an open and transparent market for energy. The day ahead market provides an opportunity for the system to commit sufficient generating units and provides financial stability for the generation suppliers while the real-time market caters to deviations and imbalances from the day ahead market.
4. **Ancillary services market** provides an open and transparent market for ancillary services. This market only takes place in the real-time time frame.

As shown in Figure 8.2, in the restructured market, the system operator administers the transmission market; the administration of the energy market can be done by the system operator or another entity.

8.2.2.3 *Present-Day Vertically Integrated Utility*   At this point in time, at least in North America, utilities have restructured in two ways:

- Full restructuring
- Restructuring with vertical integration utility model

California, for example, has adopted full restructuring. Many Canadian provinces such as British Columbia have adopted sufficient restructuring to ensure that they can offer nondiscriminatory transmission services to other entities. The structural and process changes in these utilities include

- Creating independent generation, transmission, and operation functions.
- Generation operation decisions are made by the generation division.

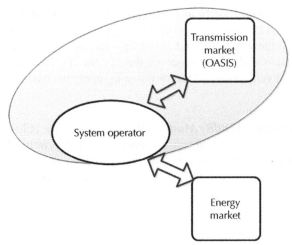

Figure 8.2   Functional entities in restructured model.

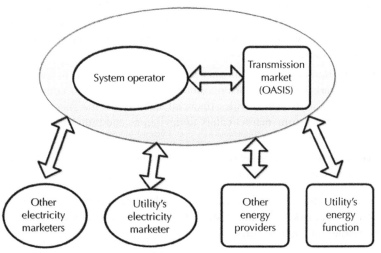

Figure 8.3   Functional entities in restructured vertically integrated utility.

- Processes are put in place for connection of independent generators for the use of the network.
- Intertie capacities are calculated and posted on the OASIS so that it can be booked by any entity on a first-come-first-serve basis.

In effect, as shown in Figure 8.3, the utility structure is modified so that there is an independent system operator operating the system and administering the transmission market. The utility's energy function competes with other energy providers in marketing and scheduling their generation. The utility could have a separate energy marketing arm that competes with other energy providers and potential energy marketers in providing and scheduling energy to the system.

8.3   PROCESS FOR GENERATION SUFFICIENCY

Regardless of the utility model, the generation sufficiency objectives and processes are similar, as they focus on ensuring that there is sufficient generation for reliably operating the system in different time frames. These processes are elaborated on in the following sections.

8.3.1   Generation Sufficiency Process for Operations Planning

Operations planning takes a longer view of system operation. The goal of operations planning is to ensure that sufficient energy resources are available within the next few months to supply the load. Operations planning considers transmission and generation outages, as well as load forecast errors. For that reason, different criteria are used to ensure generation sufficiency. There are different ways of determining sufficient reserve:

1. Simulate forecasted peak load condition contingency analysis considering $(n - 1)$ transmission and generation contingencies.
2. Use a planning reserve criterion such as
   a. The largest generating system unit
   b. One day in 10 year loss of load probability resulting in 18% of peak load [47]

One of the main reasons for the existence of this process in the restructured utility model is to establish economic conditions that encourage investment by energy providers and demand reduction resources. It is argued that the rear Real-time market and real-time market only consider the supplier's short-term marginal cost and do not reflect their total cost. This process is necessary to encourage the energy providers to make sufficient investments in the long term.

Another aspect of operations planning is to ensure that the reserve can be dispatched within an appropriate time frame. There are three different types of reserves [48]:

- *Regulating reserve:* These units consist of online fast-acting generators dispatched every 2–6 seconds.
- *Fast reserves:* These units should be available within 10 minutes; they consist of online spinning units or fast nonspinning units that can be brought online fast.
- *Slower reserve:* These units consist of offline units available within 30 minutes.

8.3.2   Generation Sufficiency Process for Near Real Time

The near real-time time frame deals with implementing the necessary changes in selecting the most economic generation portfolio (unit commitment) within a day or a few hours and the optimum set point for the units (economic dispatch) to supply that load for actual operation within a day or a few hours.

The near real-time process provides more time for posturing the system to ensure supply sufficiency and system reliability. The near real-time process has a time frame of many hours to days. During this period, operators take into account the start time and the ramping rates of the generators that need to be supplying load in real time. Operators should also ensure that they deploy the regulation reserve, slow

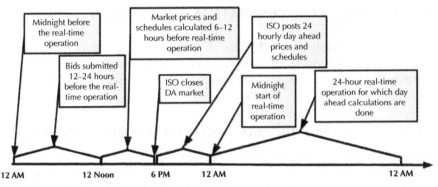

Figure 8.4   Timeline for day ahead market.

reserve, and fast reserve requirements in real-time operation to ensure that the units providing these reserves are ready for their operation in real time.

In the restructured utility model [47,48], there are normally day ahead markets for energy and ancillary services. In the day ahead market, 24-hourly prices and generation schedules are calculated for posting from midnight 18 to 24 hours ahead as per the time-line shown in Figure 8.4. As an example, the 24-hourly prices and generation schedules for November 4 are posted between 6 p.m. and 11:59 a.m. on November 3.

In a restructured vertically integrated utility, hourly generation schedules are provided by the utility's generation group as well as all the IPPs for the next day to operators.

8.3.3   Generation Sufficiency Process for Real Time

The real-time process provides opportunities for the operators to revise the near real-time schedules and posture the system to ensure supply sufficiency and system reliability. The real-time time frame ensures optimal schedules, considering the pro-jected real-time operation load, transaction schedules, transmission constraints, as well as reserve sufficiency.

In the restructured utility model, there are normally real-time markets for energy and ancillary services with intervals as short as 5–10 minutes and a shorter look-ahead horizon between 1 and 3 hours, for example.

In a restructured vertically integrated utility, operators use the most economic generation schedules to minimize the utility costs while satisfying the reserve requirements and transmission constraints.

8.4   TECHNOLOGY FOR GENERATION SUFFICIENCY

Regardless of the utility model, the generation sufficiency process uses a general optimization engine developed for power systems called "optimal power flow." (OPF). OPF forms the fundamental engine of the electricity and ancillary services markets, as well as the conventional generation portfolio optimization solution.

Generation sufficiency technology consists of two technology elements:

1. Generation sufficiency applications
2. Generation sufficiency infrastructure

8.4.1 Generation Sufficiency Applications

OPF techniques form a general mathematical tool for finding the instantaneous optimal operation of a power system under constraints, meeting the operating feasibility and security requirements [49].

These tools are used in a variety of optimization problem such as

- Generation unit commitment
- Generation dispatch
- Voltage var dispatch
- Transmission loss minimization
- Generation planning
- Production costing
- Transmission planning
- Reactive planning
- Remedial action determination

Figure 8.5 shows an EMS voltage var dispatch and transmission loss minimization display providing recommendations to the operators on the control actions he needs to take to remove violation and minimize transmission losses. For example, this display indicates that ING 2RX2 should be placed in service with a corrective reactive impact of −132.2. The last column shows the sensitivity for this control action with respect to voltage violation change and losses reduction correspondingly. Another display for this application is given in Figure 8.6. The enlarged portion of the display shows that the optimal voltage value for ING substation is 532.2 kV with the recommendation that reactor R12X4 should be placed out of service. The diagram shows that the sensitivity of the objective function with respect to voltage violation is zero because the voltage is not violated, whereas the sensitivity of the objective function with respect to system losses is 3.6.

All OPF methods are based on general optimization principles and techniques [50]. The three chief components of any static optimization processes are

- The minimum seeking approach
- The approach for handling equality constraints
- The approach for handling inequality constraints

An optimal power flow general formulation can be stated as:

Find values of u (control variables) and x (state variables) to minimize an objective function subject to the power flow equations and to the inequality constraints

Voltage-Var Dispatch Control Summary

VVD   REALTIME   VALID SOLUTION

OPF NORMAL CONVERGENCE

	Objective Function	MW Losses
Starting:	3507.62	347.20
Change:	-1.95	-2.07
Final:	3505.67	345.13

Loss Identifier

BCH
(Area or Company)

Show Controls: ▲ SELECTED
Activated Priority: Disabled

Control Type	Identifiers	Control Values BaseCase & Cost	Change & Cost ▶	Solved & Cost	Effective Limits Maximum Minimum	Response Time ▶	Priority	Control Ranking to Correct Violations ↓ Reduce Cost/Loss ▶
SHUNT	ING 2RX2	0.0	-132.2	-132.2	0.0 -132.2	2.0	0	6.99 4.30
SHUNT	MDN 2RX1	-132.2	132.2	0.0	0.0 -132.2	2.0	0	0.03 0.76
SHUNT	ACK 5RX7	-122.4	122.4	0.0	0.0 -122.4	2.0	0	1.92 0.12
SHUNT	CBK 5RX4	-122.4	122.4	0.0	0.0 -122.4	2.0	0	6.20 0.00
SHUNT	DMR 5RX4	0.0	-122.4	-122.4	0.0 -122.4	2.0	0	0.02 0.11
SHUNT	WSN 12RX2	-81.3	81.3	0.0	0.0 -81.3	2.0	0	0.00 3.13
SHUNT	ING 12RX5	-74.7	74.7	0.0	0.0 -74.7	2.0	0	0.00 3.88
SHUNT	MDN 12RX31	-35.5	35.5	0.0	0.0 -35.5	2.0	0	0.00 3.57
SHUNT	MDN 12RX32	-35.5	35.5	0.0	0.0 -35.5	2.0	0	0.00 3.57
SHUNT	UHT 3RX1	-35.0	35.0	0.0	0.0 -35.0	0.0	0	0.00 4.16
SHUNT	UHT 3RX4	-35.0	35.0	0.0	0.0 -35.0	0.0	0	0.00 4.16

Figure 8.5   Voltage var optimization display at BC Hydro.

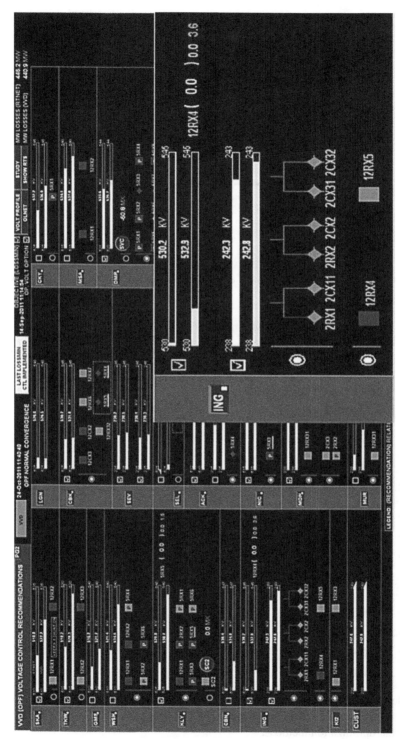

Figure 8.6   Voltage var optimization display at BC Hydro.

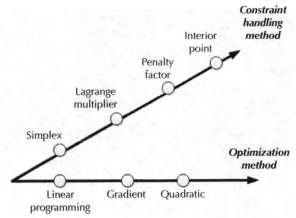

Figure 8.7 OPF methods.

that meet operation security. This is a typical nonlinear optimization problem, which can be stated mathematically as

$$\text{Minimize} \quad f(u, x) \tag{8.1}$$

$$\text{subject to} \quad h(u, x) = 0 \tag{8.2}$$

$$\text{and} \quad g(u, x) \leq 0 \tag{8.3}$$

OPF methods can be classified by two main features [51]:

1. Optimization method
2. Constraint handling solution

As shown in Figure 8.7, the main optimization methods include:

- Linear programming
- Gradient
- Quadratic

while the main constraint handling methods include:

- Simplex
- Lagrange multiplier
- Quadratic multiplier
- Interior point

While all combinations of the optimization methods and constraint handling methods are not available, this classification provides a good approach to discriminate between different optimization methods.

8.4.1.1 *OPF Optimization Techniques* This section describes different optimization techniques. In general, the OPF techniques can fall under the following categories:

1. Linear programming techniques
2. Gradient techniques
3. Quadratic techniques

These methods have been discussed in more detail in the following subsections. Another element that affects the optimization algorithm is the method for handling equality and nonequality constraints. This issue has also been addressed next.

8.4.1.2 *Linear Programming Techniques* The nonlinear OPF problem can be linearized to convert it to a linear optimization problem, which can be solved by linear programming techniques. This results in a very fast and reliable method [52,53], making it a very suitable method for the applications where the reliability and fast turnaround time is crucial.

The linear programming method, invented by George Dantzig in 1951, can be written in general as

$$\text{Minimize} \quad f(u,x) = a_1 x_1 + a_2 x_2 + \cdots a_n x_n \tag{8.4}$$

$$\text{subject to} \quad g_i(u,x) = b_{i1} x_1 + b_{i2} x_2 + \cdots b_{in} x_n \tag{8.5}$$

$$\text{and} \quad h_i(u,x) = c_{i1} x_1 + c_{i2} x_2 + \cdots c_{im} x_m \geq 0 \tag{8.6}$$

The first step in the solution of linear programming problems is to create an auxiliary problem by the introduction of artificial variables s into the inequality equations, turning them into equality as shown next:

$$h_i(u,x) = c_{i1} x_1 + c_{i2} x_2 + \cdots c_{im} x_m + s_i = 0 \tag{8.7}$$

where x_i and s_i are ≥ 0.

This method can be summarized as categorizing the equality and nonequality constraint variables as basic (free to move) or nonbasic (fixed on limit). During the optimization process, the matrices for basic and nonbasic variables are continuously updated.

The steps in solving a traditional simplex method can be summarized as

- Start with selecting all xi variables as nonbasic variables assigning their values as zero.
- Start with selecting all si variables as basic variables assigning different values to them to convert the inequality constraints to equality constraints.
- Try to change one x_i variable value from zero to another value to minimize the objective function $f(u, x)$, swapping it with a nonbasic variable s_i.
- Choose the x_i with highest impact (e.g., impact on minimizing or maximizing the objective function).

- Choose the basic variable that approaches zero first. This can be obtained by retaining in the constraint equations only the x_i variable that we want to swap and use the actual values obtained for other x_i variables. All the s_i variables are also retained. This is followed up by getting the x_i values for which s_i variables become zero. The si corresponding to the highest x_i value will be chosen for swapping.
- The objective function will be written in terms of s_i that has come in as zero.
- Continue until no further reduction is possible.

Example 8.1
Minimize the following function using the simplex method:

$$f(x) = -x_1 - 1.5x_2 - x_3$$

Subject to

$$4x_1 - 2x_2 + x_3 \leq 5$$
$$2x_1 + 3x_2 - 8x_3 \leq 2.5$$
$$1.5x_1 + 4x_2 + 3x_3 \leq 7$$

Solution
We first develop equality constraints by introducing basic variables s_1, s_2, and s_3. Please note that all nonbasic and basic variables have positive values.

Iteration 1

$$\begin{aligned}
s_1 &: 4x_1 - 2x_2 + x_3 + s_1 = 5 \\
s_2 &: 2x_1 + 3x_2 - 8x_3 + s_2 = 2.5 \\
s_3 &: 1.5x_1 + 4x_2 + 3x_3 + s_3 = 7 \\
\pi &= -x_1 - 1.5x_2 - x_3
\end{aligned} \tag{8.8}$$

x_2 has the most effect on the objective function because it has the largest negative coefficient in the objective function. So set every variable except for x_2 into (8.8) and note the basic variable in the equations:

$$\begin{aligned}
s_1 &: -2x_2 = 5 \rightarrow x_3 = -2.5 \\
s_2 &: 3x_2 = 2.5 \rightarrow x_3 = 0.8333 \\
s_3 &: 4x_2 = 7 \rightarrow x_3 = 1.75 \\
\pi &= -1.5x_3
\end{aligned} \tag{8.9}$$

Find the smallest nonnegative x_3 when basic variables have been reduced to zero and note the slack variable s_2. Now solve for x_2 in (8.8):

$$\begin{aligned}
s_2 &: x_1 + 3x_2 - 8x_3 + s_2 = 2.5 \\
3x_2 &= -x_1 + 8x_3 - s_2 + 2.5 \\
x_2 &= -0.6667x_1 + 2.6667x_3 - 0.3333s_2 + 0.8333
\end{aligned} \tag{8.10}$$

Substitute the value of x_2 from (8.10) into (8.8):

$$s_1 : 4x_1 - 2(-0.6667x_1 + 2.6667x_3 - 0.3333s_2 + 0.8333) + x_3 + s_1 = 5$$
$$x_2 : 0.6667x_1 + x_2 - 2.6667x_3 + 0.3333s_2 = 0.8333$$
$$s_3 : 1.5x_1 + 4(-0.6667x_1 + 2.6667x_3 - 0.3333s_2 + 0.8333) + 3x_3 + s_3 = 7 \tag{8.11}$$
$$\pi = -x_1 - 1.5(-0.6667x_1 + 2.6667x_3 - 0.3333s_2 + 0.8333) - x_3$$

Simplifying equations given in (8.11) results in

$$s_1 : 5.3333x_1 - 4.3333x_3 + 0.6667s_2 + s_1 = 6.7777$$
$$x_2 : 0.6667x_1 + x_2 - 2.6667x_3 + 0.3333s_2 = 0.8333$$
$$s_3 : -1.167x_1 + 13.67x_3 - 1.3333s_2 + s_3 = 3.667 \tag{8.12}$$
$$\pi = -5x_3 + 0.5s_2 - 1.25$$

Iteration 2

x_3 has the most effect on the objective function because it has the largest negative coefficient. So set every nonbasic variable except for x_3 to zero in (8.12):

$$s_1 : -4.3333x_3 = 6.7777 \rightarrow x_3 = -1.5641$$
$$x_2 : -2.6667x_3 = 0.8333 \rightarrow x_3 = -0.3125$$
$$s_3 : 13.67x_3 = 3.667 \rightarrow x_3 = 0.23825 \tag{8.13}$$
$$\pi = -5x_3$$

Find the smallest nonnegative x_3 and note the slack variable s_3. Now rewriting Equation 8.12 in terms of s_3 instead of x_3 gives

$$s_3 : -1.167x_1 + 13.67x_3 - 1.3333s_2 + s_3 = 3.667$$
$$s_3 : 13.67x_3 = 1.167x_1 + 1.3333s_2 - s_3 + 3.667 \tag{8.14}$$
$$s_3 : x_3 = 0.08637x_1 + 0.09756s_2 - 0.07317s_3 + 0.26823$$

Substitute Equation 8.14 into Equation 8.12:

$$s_1 : 5.3333x_1 - 4.3333(0.08637x_1 + 0.09756s_2 - 0.07317s_3 + 0.26823)$$
$$+ 0.6667s_2 + s_1 = 6.7777$$
$$x_2 : 0.6667x_1 + x_2 - 2.6667(0.08637x_1 + 0.09756s_2 - 0.07317s_3 + 0.26823) \tag{8.15}$$
$$+ 0.3333s_2 = 0.8333$$
$$x_3 : -0.08637x_1 + x_3 - 0.09756s_2 - 0.07317s_3 = 0.26823$$
$$\pi = -5(0.08637x_1 + 0.09756s_2 - 0.07317s_3 + 0.26823) + 0.5s_2 - 1.25$$

Simplifying equations given in (8.15) results in

$$s_1 : 4.963x_1 + s_1 + 0.2439s_2 + 0.3171s_3 = 7.829$$
$$x_2 : 0.439x_1 + x_2 + 0.07317s_2 + 0.1951s_3 = 0.8333$$
$$x_3 : -0.08637x_1 + x_3 - 0.09756s_2 - 0.07317s_3 = 0.26823 \tag{8.16}$$
$$\pi = -0.4268x_1 + 0.0122s_2 + 0.3659s_3 - 2.591$$

Iteration 3

x_1 has the most effect on the objective function because it has the largest negative coefficient. So set every variable except for x_1 into (8.16) and note the slack variable in the equation:

$$
\begin{aligned}
s_1 &: 4.963x_1 = 7.829 \rightarrow x_1 = 1.5775 \\
x_2 &: 0.439x_1 = 0.8333 \rightarrow x_1 = 1.898 \\
x_3 &: -0.08637x_1 = 0.26823 \rightarrow x_1 = -3.1056 \\
\pi &= -0.4268x_1
\end{aligned}
\tag{8.17}
$$

Find the smallest nonnegative x_1 and note the slack variable s_1. Now solve for x_1 in (8.23):

$$
\begin{aligned}
s_1 &: 4.963x_1 + s_1 + 0.2439s_2 + 0.3171s_3 = 7.829 \\
s_1 &: 4.963x_1 = -s_1 - 0.2439s_2 - 0.3171s_3 + 7.829 \\
s_1 &: x_1 = -0.2015s_1 - 0.04914s_2 - 0.06388s_3 + 1.577
\end{aligned}
\tag{8.18}
$$

Substitute Equation 8.18 into Equation 8.16:

$$
\begin{aligned}
x_1 &: x_1 + 0.2015s_1 + 0.04914s_2 + 0.06388s_3 = 1.577 \\
x_2 &: 0.439(-0.2015s_1 - 0.04914s_2 - 0.06388s_3 + 1.577) \\
    &\quad + x_2 + 0.07317s_2 + 0.1951s_3 = 0.8333 \\
x_3 &: -0.08637(-0.2015s_1 - 0.04914s_2 - 0.06388s_3 + 1.577) \\
    &\quad + x_3 - 0.09756s_2 - 0.07317s_3 = 0.26823 \\
\pi &= -0.4268(-0.2015s_1 - 0.04914s_2 - 0.06388s_3 + 1.577) \\
    &\quad + 0.0122s_2 + 0.3659s_3 - 2.591
\end{aligned}
\tag{8.19}
$$

Simplifying equations given in (8.19) results in

$$
\begin{aligned}
x_1 &: x_1 + 0.2015s_1 + 0.04914s_2 + 0.06388s_3 = 1.577 \\
x_2 &: x_2 - 0.08845s_1 + 0.0516s_2 + 0.1671s_3 = 0.8563 \\
x_3 x_3 &+ 0.172s_1 - 0.0933s_2 + 0.07862s_3 = 0.4029 \\
\pi &= 0.086s_1 + 0.03317s_2 + 0.3931s_3 - 3.265
\end{aligned}
\tag{8.20}
$$

There are now further reductions possible; therefore, x_1, x_2, x_3 are the basic variables and take the values

$$
x_1 = 1.577, \quad x_2 = 0.8563, \quad x_3 = 0.4029 \quad \therefore \pi = -3.265
$$

8.4.1.3 *Gradient Techniques*

The basic principle of the gradient technique [50,54] is to express the gradient of the objective function as a function of selected independent variables. For each variable, a progression direction is then defined as the opposite of the reduced gradient. If this progression direction is a zero vector, the optimum is reached; otherwise, the progression is carried out until the objective function decreases and a new constraint is met. These methods by nature have not been very inefficient for use in large-scale applications.

The gradient method can be derived from the Taylor expansion of $f(x)$ around point x_0:

$$f(x) = f(x_0) + (x - x_0)f'(x_0) + (1/2)(x - x_0)^2 f''(x_0) + \cdots \qquad (8.21)$$

By only considering the first two terms, we get

$$f(x) = f(x_0) + (x - x_0)f'(x_0) \qquad (8.22)$$

The gradient method is based on the notion that $f(x)$ increases in the direction of the derivative at point x_0. So the gradient method works based on the notion that to minimize the function we need to move in the negative direction of the gradient vector:

$$x = x_0 - \alpha \cdot f'(x_0) \qquad (8.23)$$

where α is an acceleration factor and $f'(x_0)$ is the gradient of the function, which is also called "Jacobian" when x is a vector. The method is continued iteratively until the gradient becomes smaller than a threshold.

There are two ways to implement this method:

- Use a constant value for α
- Obtain an optimum value for α in each iteration. In this method, x and $f(x)$ are obtained as a function of α. Then, an optimum α is obtained, which minimizes and $f(x)$ is minimized to get the optimum.
- There are two ways to implement this method:

These two methods are described in the following examples.

Example 8.2

Minimize $y = x^2$ using the gradient method starting at $(1, 1)$.

Method 1: Constant Acceleration Factor
First, we calculate the gradient of y:

$$y' = 2x$$

We then calculate the correction to x:

$$\Delta x = -\alpha \frac{df}{dx}$$

Let us first assume that we use a constant value for α such as $\alpha = 0.4$. Then we get the following corrections:

$$\Delta x(1) = -0.8, x_1 = 0.2$$

$$\Delta x(2) = -0.16, x_2 = 0.04$$

$$\Delta x(3) = -0.032, x_3 = 0.008$$

which converges to $x = 0$ in a few iterations.

Method 2: Optimal Acceleration Factor
Now we repeat the same solution with an optimal value in each iteration:
Recall that $y' = 2x$ and $\Delta x = -\alpha df/dx$. Therefore,

$$\Delta x(1) = -2\alpha$$

which gives

$$x_1 = (1-2\alpha) \quad \text{and} \quad y = (1-2\alpha)^2$$

We now obtain a value for α, which minimizes y:

$$\frac{dy}{d\alpha} = 0 = (1-2\alpha) = 0$$

$$\alpha = 0.5$$

So with this value of α we get

$$\Delta x(1) = -(0.5)(2) = -1, \ x_1 = 1 + (-1) = 0$$

We can now repeat the process calculating the correction $\Delta x(2)$:

$$\Delta x(2) = -2 * 0 = 0$$

indicating that the problem has converged to the solution in one iteration.

Example 8.3
Minimize the following function using the gradient method

$$f(x_1, x_2) = 2x_1^2 + 9x_2^2 + 2x_1 x_2 - 8x_1 + 3x_2$$

with the values starting at

$$\begin{bmatrix} x_1^i \\ x_2^i \end{bmatrix} = \begin{bmatrix} x_1^0 \\ x_2^0 \end{bmatrix} = \begin{bmatrix} 3 \\ 5 \end{bmatrix}$$

Solution
We first develop the gradient of function f:

$$\nabla f(x_1, x_2) = \begin{bmatrix} \dfrac{\partial f}{\partial x_1} \\ \dfrac{\partial f}{\partial x_2} \end{bmatrix} = \begin{bmatrix} 4x_1 + 2x_2 - 8 \\ 18x_2 + 2x_1 + 3 \end{bmatrix}$$

$$f\begin{pmatrix} x_1^1 \\ x_2^1 \end{pmatrix} = 264$$

Iteration 1

New values are calculated from using the gradients with respect to x_1 and x_2:

$$\nabla f(3,5) = \begin{bmatrix} 4(3) + 2(5) - 8 \\ 18(5) + 2(3) + 3 \end{bmatrix} = \begin{bmatrix} 14 \\ 99 \end{bmatrix}$$

$$\begin{bmatrix} x_1^{i+1} \\ x_2^{i+1} \end{bmatrix} = \begin{bmatrix} x_1^i \\ x_2^i \end{bmatrix} - \gamma_i \nabla f(3,5)$$

$$\begin{bmatrix} x_1^2 \\ x_2^2 \end{bmatrix} = \begin{bmatrix} 3 \\ 5 \end{bmatrix} - \gamma_i \begin{bmatrix} 14 \\ 99 \end{bmatrix}$$

$$f\left(x_1^2 = 3 - 14 \times \gamma_0, x_2^2 = 5 - 99 \times \gamma_0\right) = 91,373.0 \times \gamma_0^2 - 9,997.0 \times \gamma_0 + 264$$

$$\frac{\partial f}{\partial \gamma_0} = 182,746 \times \gamma_0 - 9,997 = 0 \rightarrow \gamma_0 = 0.055$$

$$\begin{bmatrix} x_1^2 \\ x_2^2 \end{bmatrix} = \begin{bmatrix} 1 \\ 2 \end{bmatrix} - 0.055 \times \begin{bmatrix} 14 \\ 99 \end{bmatrix} = \begin{bmatrix} 2.23 \\ -0.42 \end{bmatrix}$$

$$f\begin{pmatrix} x_1^2 \\ x_2^2 \end{pmatrix} = -9.44$$

Before moving to the next iteration, we need to test the actual change in the value of function f:

$$\left(\frac{f\begin{pmatrix} x_1^{i+1} \\ x_2^{i+1} \end{pmatrix} - f\begin{pmatrix} x_1^i \\ x_2^i \end{pmatrix}}{f\begin{pmatrix} x_1^i \\ x_2^i \end{pmatrix}} < \text{Change threshold} \right) \text{then} \begin{cases} \text{if yes, then Stop Iterations} \\ \text{if no, then Continue} \end{cases}$$

$$\text{Change in } f = \left(\frac{264 - (9.44)}{264} \right) = 1.04$$

Since the change in function f is large, we continue to the second iteration:

Iteration 2: $i = 2$

$$\nabla f\left(\begin{bmatrix} 2.23 \\ -0.42 \end{bmatrix} \right) = \begin{bmatrix} 12 \times 2.23 + 4 \times -0.42 - 85 \\ 6 \times -0.42 + 4 \times 2.23 + 3 \end{bmatrix} = \begin{bmatrix} 0.105 \\ -0.015 \end{bmatrix}$$

$$\begin{bmatrix} x_1^{i+1} \\ x_2^{i+1} \end{bmatrix} = \begin{bmatrix} x_1^i \\ x_2^i \end{bmatrix} - \gamma_i \nabla f(1,2)$$

$$\begin{bmatrix} x_1^3 \\ x_2^3 \end{bmatrix} = \begin{bmatrix} 2.23 \\ -0.42 \end{bmatrix} - \gamma_i \begin{bmatrix} 0.105 \\ -0.015 \end{bmatrix}$$

$$f\left(x_1^3 = 2.23 - 0.11 \times \gamma_0, x_2^3 = -0.42 + 0.015 \times \gamma_0\right) = 0.021 \times \gamma_0^2 - 0.011 \times \gamma_0 - 9.44$$

$$\frac{\partial f}{\partial \gamma_0} = 0.042 \times \gamma_0 - 0.011 = 0 \rightarrow \gamma_0 = 0.27$$

$$\begin{bmatrix} x_1^3 \\ x_2^3 \end{bmatrix} = \begin{bmatrix} 2.23 \\ -0.42 \end{bmatrix} - 0.27 \begin{bmatrix} 0.105 \\ -0.015 \end{bmatrix} = \begin{bmatrix} 2.21 \\ -0.41 \end{bmatrix}$$

$$f\begin{pmatrix} x_1^3 \\ x_2^3 \end{pmatrix} = -9.44$$

$$\left(\frac{f\begin{pmatrix} x_1^{i+1} \\ x_2^{i+1} \end{pmatrix} - f\begin{pmatrix} x_1^i \\ x_2^i \end{pmatrix}}{f\begin{pmatrix} x_1^i \\ x_2^i \end{pmatrix}} < \text{Change threshold} \right) \text{then} \begin{cases} \text{if yes, then Stop Iterations} \\ \text{if no, then Continue} \end{cases}$$

It is evident that

$$f\begin{pmatrix} x_1^2 \\ x_2^2 \end{pmatrix} \sim f\begin{pmatrix} x_1^3 \\ x_2^3 \end{pmatrix}$$

rendering the change in $f = 0$ stopping the processes. We can test the values obtained by recalculating the gradient function:

$$\nabla f(x_1, x_2) = \begin{bmatrix} 4x_1 + 2x_2 - 8 \\ 18x_2 + 2x_1 + 3 \end{bmatrix} = \begin{bmatrix} 0 \\ 0 \end{bmatrix}$$

$$\begin{bmatrix} 4 & 2 \\ 2 & 18 \end{bmatrix} \begin{bmatrix} x_1 \\ x_2 \end{bmatrix} = \begin{bmatrix} 8 \\ -3 \end{bmatrix}$$

$$\begin{bmatrix} x_1 \\ x_2 \end{bmatrix} = \begin{bmatrix} 4 & 2 \\ 2 & 18 \end{bmatrix}^{-1} \begin{bmatrix} 8 \\ -3 \end{bmatrix} = \begin{bmatrix} 2.21 \\ -0.41 \end{bmatrix}$$

$$f\left(\begin{bmatrix} 2.21 \\ -0.41 \end{bmatrix}\right) = -9.44$$

8.4.1.4   Quadratic Techniques   The basic idea of quadratic techniques [55–57] is to set the value of the gradient $\nabla f(x)$ of the convex objective function $f(x)$ zrero at the optimum. We now use Taylor to expand $f(x)'$ around x_0:

$$f(x) = f'(x_0) + (x - x_0)f''(x_0) + (1/2)(x - x_0)^2 f'''(x_0) + \cdots \tag{8.24}$$

Using only the linear term, we get

$$f'(x) - f'(x_0) + (x - x_0)f''(x_0) = 0 \tag{8.25}$$

In multivariable systems, the first derivative is called Jacobian and the second derivative is called Hessian. This equation can be written in the form

$$f'(x) - J_0 + \Delta x H_0 = 0 \tag{8.26}$$

where $\Delta x = (x - x_0)$ and J_0 and H_0 are Jacobian n and Hessian values at point 0. This equation is solved iteratively by finding a correction Δx that sets $f'(x) = 0$ giving

$$\Delta x = -\left(\frac{J_0}{H_0}\right) \tag{8.27}$$

$$x_1 = x_0 + (\Delta x) \tag{8.28}$$

and the process continues until Δx becomes smaller than a threshold.

Example 8.4
Minimize $f(x) = x^2$ using the Newton method at $(1, 1)$.

$$y' = 2x, \quad y'' = 2$$

Iteration 1: $(x = 1)$

$$\Delta x = -\frac{y'}{y''} = -1$$

$$\Delta x_1 = -\frac{2(1)}{2} = -1$$

$$x_1 = -1 + 1 = 0 \quad \text{and} \quad f(x) = 0$$

Iteration 2: $(x = 0)$

$$\Delta x_2 = \frac{2(0)}{2} = 0$$

Since $\Delta x = 0$, we know that the solution has converged; the function is minimum at $x = 0$.

Example 8.5

Minimize the following function using Newton's method:

$$f(x_1, x_2) = 6x_1^2 + 3x_2^2 + 4x_1 x_2 + 5.5x_1 - 2x_2$$

starting at the initial values of

$$\begin{bmatrix} x_1^i \\ x_2^i \end{bmatrix} = \begin{bmatrix} x_1^0 \\ x_2^0 \end{bmatrix} = \begin{bmatrix} 1 \\ 2 \end{bmatrix}$$

Solution

$$\nabla f(x_1, x_2) = \begin{bmatrix} \dfrac{\partial f}{x_1} \\ \dfrac{\partial f}{x_2} \end{bmatrix} = \begin{bmatrix} 4x_1 + 2x_2 - 8 \\ 18x_2 + 2x_1 + 3 \end{bmatrix}$$

$$H = \begin{bmatrix} 4 & 2 \\ 2 & 18 \end{bmatrix}$$

$$H^{-1} = \begin{bmatrix} 0.264706 & -0.029412 \\ -0.029412 & 0.058824 \end{bmatrix}$$

Iteration 1

$$\nabla f(3,5) = \begin{bmatrix} 14 \\ 99 \end{bmatrix}$$

$$-H^{-1}\nabla f(3,5) = \begin{bmatrix} -0.794118 \\ -5.41176 \end{bmatrix}$$

$$\begin{bmatrix} x_1^2 \\ x_2^2 \end{bmatrix} = \begin{bmatrix} 1 \\ 2 \end{bmatrix} + \begin{bmatrix} -0.794118 \\ -5.41176 \end{bmatrix} = \begin{bmatrix} 2.20588 \\ -0.411765 \end{bmatrix}$$

Iteration 2

$$\nabla f\left(\begin{bmatrix} 2.20588 \\ -0.411765 \end{bmatrix} \right) = \begin{bmatrix} 0 \\ 0 \end{bmatrix}$$

$$-H^{-1}\nabla f(2.20588, -0.411765) = \begin{bmatrix} 0 \\ 0 \end{bmatrix}$$

Since the gradient is zero, we stop the iteration process.

8.4.1.5 *Constraint Handing Methods*

Karush–Kuhn–Tucker [52–58] is the general method providing sufficient condition to solve a constrained optimization problem. The method basically converts a constrained optimization problem given in Equations 8.1–8.3 into a nonconstrained problem by adding the constraints into the objective function. The method defines a new objective function called Lagrangian as follows:

$$L = f(x,u) + \lambda h(x,u) + \beta a(x,z,u) \tag{8.29}$$

where $h(x, u)$ is the equality constraint and $a(x, z, u)$ is the equality constraint resulting from the introduction of variable z to the inequality constraint $g(x, u) \le 0$:

$$a(x,z,u) = g(x,u) + z^2 = 0 \tag{8.30}$$

Using a z^2 instead of z makes the solution insensitive to whether z is negative or positive. Please note that if the inequality is in the form of $g(x,u) \ge 0$, then the resulting equality equation becomes

$$a(x,z,u) = g(x,u) - z^2 = 0 \tag{8.31}$$

Karush–Kuhn–Tucker sufficient conditions are

$$\begin{aligned}
\frac{\partial L}{\partial x} &= 0 \\
\frac{\partial L}{\partial z} &= 0 \\
\frac{\partial L}{\partial \lambda} &= 0 \\
\frac{\partial L}{\partial \beta} &= 0
\end{aligned} \tag{8.32}$$

This method can be applied in different ways as given next.

1. Lagrange Multiplier Method

This is the exact implementation given in Equation 8.16 using Lagrange multipliers β and λ. The main drawback of this method is the need to identify the correct set of binding constraints in each iteration. In each iteration, the inequality constraints need to be examined, and those that are binding (in violation) need to be added to the active set. A few methods have been proposed for rectifying this problem [57].

Example 8.6

Minimize $f(x) = (x)^2$

　　　Subject to $g(x) = x - 1 \ge 0$ using Kuhn–Tucker method with Lagrange multiplier constraint handling approach.

Solution

We develop the Lagrangian function:

$$£ = f(x) + \beta \cdot (x - 1 + z^2)$$

$$\frac{\partial £}{\partial x} = 0 \quad 2x + \beta = 0 \Rightarrow x = -\frac{\beta}{2}$$

$$\frac{\partial £}{\partial \beta} = 0 \quad x - 1 + z^2 = 0 \Rightarrow x = 1$$

$$\frac{\partial £}{\partial z} = 0 \quad 2\beta z = 0$$

From the last equation, we see either $\beta = 0$ or $z = 0$. With $\beta = 0$, we get $x = 0$, violating the constraint. On the other hand, with $z = 0$, we get $x = 1$, which is the acceptable solution.

Example 8.7

Minimize the following function using the Kuhn–Tucker method with Lagrange multiplier constraint handling approach:

$$f(x) = 6x_1^2 + 3x_2^2$$

subject to

$$g_1(x) = -x_1 + x_2 \geq 0.4$$

Solution

First, we change the inequality constraints to the following form

$$g_1(x) = x_1 - x_2 \leq -0.4$$

Then we transform the inequality to equality by adding a slack variable x_3^2:

$$h_1(x) = x_1 - x_2 + x_3^2 + 0.4 = 0$$

$$L = 6x_1^2 + 3x_2^2 + \lambda_1 \left(x_1 - x_2 + x_3^2 + 0.4 \right)$$

We then develop the first-order derivatives of Lagrangian with respect to x_1, x_2, and x_3:

$$\nabla L = \begin{bmatrix} 12x_1 + \lambda_1 \\ 6x_2 - \lambda_1 \\ 2\lambda_1 x_3 \end{bmatrix}$$

Then we develop the second-order derivatives of Lagrangian with respect to x_1, x_2, and x_3:

$$\nabla^2 L = \begin{bmatrix} 12 & 0 & 0 \\ 0 & 6 & 0 \\ 0 & 0 & 2\lambda_1 \end{bmatrix}$$

We develop the derivative of the equality constraint $h_1(x)$ with respect to x_1, x_2, and x_3:

$$G = \nabla h_1(x) = \begin{bmatrix} 1 \\ -1 \\ 2x_3 \end{bmatrix}$$

The developed matrices $\nabla^2 L$ and G are used to develop the Jacobian of J, which is the second derivative of L with respect to x_1, x_2, x_3, and λ_1. This will be used in the Newton Equation 8.14 to find the values of x_1, x_2, x_3, and λ_1 for which the solution is minimum:

$$J = \begin{bmatrix} 12 & 0 & 0 & 1 \\ 0 & 6 & 0 & -1 \\ 0 & 0 & 2\lambda_1 & 2x_3 \\ 1 & -1 & 2x_3 & 0 \end{bmatrix} = \begin{bmatrix} \nabla^2 L & G \\ G^T & 0 \end{bmatrix}$$

Solve using these steps:

$$\begin{bmatrix} \nabla^2 L & G \\ G^T & 0 \end{bmatrix}^j \begin{bmatrix} \Delta x^j \\ \Delta \lambda^j \end{bmatrix} = -\begin{bmatrix} \nabla L \\ h \end{bmatrix}^j$$

$$\begin{bmatrix} \Delta x^j \\ \Delta \lambda^j \end{bmatrix} = -\left(\begin{bmatrix} \nabla^2 L & G \\ G^T & 0 \end{bmatrix}^j \right)^{-1} \begin{bmatrix} \nabla L \\ h \end{bmatrix}^j$$

$$\begin{bmatrix} x^{j+1} \\ \lambda^{j+1} \end{bmatrix} = \begin{bmatrix} x^j \\ \lambda^j \end{bmatrix} + \begin{bmatrix} \Delta x^j \\ \Delta \lambda^j \end{bmatrix}$$

Iteration 1: $j = 1$

$$\begin{bmatrix} x^0 \\ \lambda^0 \end{bmatrix} = \begin{bmatrix} 1 \\ 1 \\ 0 \\ 1 \end{bmatrix}$$

$$f(x^j) = 6x_1^2 + 3x_2^2 = 9$$

$$g_1(x^j) = -x_1 + x_2 = 0$$

$$\begin{bmatrix} \nabla L \\ h \end{bmatrix}^j = \begin{bmatrix} 11 \\ 7 \\ 0 \\ 0.4 \end{bmatrix}$$

$$\begin{bmatrix} \nabla^2 L & G \\ G^{\mathrm{T}} & 0 \end{bmatrix}^j = \begin{bmatrix} 12 & 0 & 0 & 1 \\ 0 & 6 & 0 & -1 \\ 0 & 0 & 2 & 0 \\ 1 & -1 & 0 & 0 \end{bmatrix}$$

$$\begin{bmatrix} \Delta x^j \\ \Delta \lambda^j \end{bmatrix} = -\left(\begin{bmatrix} \nabla^2 L & G \\ G^{\mathrm{T}} & 0 \end{bmatrix}^j\right)^{-1}\begin{bmatrix} \nabla L \\ h \end{bmatrix}^j = \begin{bmatrix} -1.1333 \\ -0.7333 \\ 0 \\ 0.6 \end{bmatrix}$$

$$\begin{bmatrix} x^{j+1} \\ \lambda^{j+1} \end{bmatrix} = \begin{bmatrix} x^j \\ \lambda^j \end{bmatrix} + \begin{bmatrix} \Delta x^j \\ \Delta \lambda^j \end{bmatrix} = \begin{bmatrix} 1 \\ 1 \\ 0 \\ 1 \end{bmatrix} + \begin{bmatrix} -1.1333 \\ -0.7333 \\ 0 \\ 0.6 \end{bmatrix} = \begin{bmatrix} -0.1333 \\ 0.2667 \\ 0 \\ 1.6 \end{bmatrix}$$

$$f(x^{j+1}) = 6x_1^2 + 3x_2^2 = 0.32$$

$$g_1(x^{j+1}) = -x_1 + x_2 = 0.4$$

Iteration 2: $j = 2$

$$\begin{bmatrix} \nabla L \\ h \end{bmatrix}^j = \begin{bmatrix} 0 \\ 0 \\ 0 \\ 0 \end{bmatrix}$$

indicating that the solution has converged.

Table 8.1 gives the details of the calculations. Figure 8.8 shows the optimization problem of Example 8.7 as a function of x_1 and x_2 with the acceptable optimization region shown in a darker shade.

2. Penalty Function Method

A variation to the Lagrangian by adding penalties for the constraints mostly in the form of quadratic functions [59].

$$L = f(x,u) + \lambda h(x,u) + \beta a^2(x,u) \tag{8.33}$$

The penalty factor is increased in each iteration by a factor β (e.g., 10) to speed up the convergence. Again similar to the Lagrange multiplier method, the main drawback of this method is the need to identify the correct set of binding constraints in each iteration [57].

Example 8.8

Minimize $f(x) = (x)^2$

Subject to $g(x) = x - 1 \geq 0$ using the Kuhn–Tucker method with penalty function constraint handling approach.

TABLE 8.1 Calculation details of Example 8.7

J	$\begin{bmatrix} x^j \\ \lambda^j \end{bmatrix}$	$\begin{bmatrix} \nabla^2 L & G^j \\ G^T & 0 \end{bmatrix}$				$\left(\begin{bmatrix} \nabla^2 L & G^j \\ G^T & 0 \end{bmatrix}\right)^{-1}$				$\begin{bmatrix} \nabla L \\ h \end{bmatrix}$	$\begin{bmatrix} \Delta x^j \\ \Delta \lambda^j \end{bmatrix}$	$\begin{bmatrix} x^{j+1} \\ \lambda^{j+1} \end{bmatrix}$
1	1	12	0	0	-1	0.055556	0.055556	0	-0.33333	11	1.133333	-0.13333
	1	0	6	0	1	0.055556	0.055556	0	0.666667	7	-0.733333	0.266667
	0	0	0	2	0	0	0	0.5	0	0	0	0
	1	-1	1	0	0	-0.33333	0.666667	0	-4	0.4	0.6	1.6
2	-0.133	12	0	0	-1	0.055556	0.055556	0	-0.33333	0	0	-0.13333
	0.267	0	6	0	1	0.055556	0.055556	0	0.666667	0	0	0.266667
	0	0	0	-3.2	0	0	0	-0.3125	0	0	0	0
	-1.6	-1	1	0	0	-0.33333	0.666667	0	-4	0	0	1.6

143

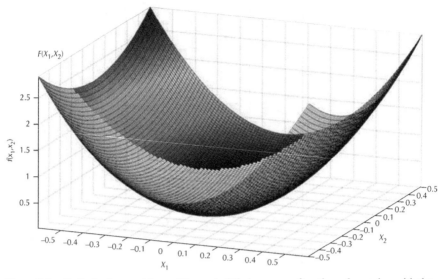

Figure 8.8   Optimization problem of Example 8.7 shown as a function of x_1 and x_2 with the acceptable optimization region shown in a darker shade.

Solution

We develop the Lagrangian function:

$$L = x^2 + \sigma(x-1)^2$$

$$\frac{\partial L}{\partial x} = 2x + 2\sigma \quad x - 2\sigma = 0$$

$$x = \frac{\sigma}{(1+\sigma)}$$

So to solve this problem, we start with a value of $\sigma = 1$ and increase it by a factor of 10 in each iteration.

Iteration 1: $\sigma = 1$, $x = 0.5$
Iteration 2: $\sigma = 10$, $x = 0.90$
Iteration 3: $\sigma = 100$, $x = 0.99$
Iteration 4: $\sigma = 1000$, $x = 0.999$

We see that the solution is converging to $x = 1$.

Example 8.9

Minimize the following function using the Kuhn–Tucker method with penalty function constraint handling approach:

$$f(x) = 6x_1^2 + 3x_2^2$$

subject to

$$g_1(x) = -x_1 + x_2 \geq 0.4$$

TABLE 8.2   Calculation details of Example 8.9

n	σ	x_n	$\nabla^2 L$		$(\nabla^2 L)^{-1}$		∇L	$(\nabla^2 L)^{-1}\nabla L$	x_{n+1}
0	1	1	14	−2	0.07407	0.01852	12.8	1.0444	−0.0444
		1	−2	8	0.01852	0.12963	5.2	0.9111	0.0889
1	10	−0.0444	32	−20	0.06019	0.04630	4.8	0.0667	−0.1111
		0.0889	−20	26	0.04630	0.07407	−4.8	−0.1333	0.2222
2	10^2	−0.1111	212	−200	0.05610	0.05447	12	0.0196	−0.1307
		0.2222	−200	206	0.05447	0.05773	−12	−0.0392	0.2614
3	10^3	−0.1307	2,012	−2,000	0.05561	0.05544	14.11	0.0023	−0.1331
		0.2614	−2,000	2,006	0.05544	0.05578	−14.11	−0.0047	0.2661
4	10^4	−0.1331	20,012	−20,000	0.05556	0.05554	14.37	0.0002	−0.1333
		0.2661	−20,000	20,006	0.05554	0.05558	−14.37	−0.0005	0.2666
5	10^5	−0.1333	200,012	−200,000	0.05556	0.05555	14.39	0.0000	−0.1333
		0.2666	−200,000	200,006	0.05555	0.05556	−14.39	0.0000	0.2667
6	10^6	−0.1333	2,000,012	−2,000,000	0.05556	0.05556	14.39	0.0000	−0.1333
		0.2667	−2,000,000	2,000,006	0.05556	0.05556	−14.39	0.0000	0.2667

Solution

$$g(x) = -x_1 + x_2 - 0.4$$

$$L = f(x) + \sigma g(x)^2$$

$$L = 6x_1^2 + 3x_2^2 + \sigma(-x_1 + x_2 - 0.4)^2$$

$$\nabla L = \begin{bmatrix} 12x_1 - 2\sigma(-x_1 + x_2 - 0.4) \\ 6x_2 + 2\sigma(-x_1 + x_2 - 0.4) \end{bmatrix}$$

$$\nabla^2 L = \begin{bmatrix} 12 + 2\sigma & -2\sigma \\ -2\sigma & 6 + 2\sigma \end{bmatrix}$$

$$x_{n+1} = x_n - (\nabla^2 L)^{-1}\nabla L$$

$$x_{n+1} = x_n - \left(\begin{bmatrix} 12 + 2\sigma & -2\sigma \\ -2\sigma & 6 + 2\sigma \end{bmatrix} \right)^{-1} \begin{bmatrix} 12x_1 - 2\sigma(-x_1 + x_2 - 0.4) \\ 6x_2 + 2\sigma(-x_1 + x_2 - 0.4) \end{bmatrix}$$

We start with the initial conditions of $x_1 = 1$, $x_2 = 1$, and $\sigma = 1$. Table 8.2 provides the calculation details.

3. Simplex-Based Methods

The description of this method is provided in Section 8.4.1.2.

4. Interior Point Methods

The interior point method has been applied to linear programming and to nonlinear optimization. This method handles the inequality constraints through logarithm barrier functions instead of penalties [60, 61, 62, 63].

So for the original problem given in Equations 8.1–8.3, using the logarithmic function to represent the inequality constraints will change the Lagrangian function to:

$$L = f(x,u) + \lambda h(x,u) + \mu \ln(g(x,u)) \tag{8.34}$$

With this approach, the main drawback of other constraint handling methods for non-linear solutions that are active set identification (set of binding inequality constraint) is eliminated. It must be noted that Equation 8.20 can only work for $g(x, u) \geq 0$. So the constraints of the form $g(x, u) \leq 0$ should be changed to $-g(x, u) \geq 0$. The barrier factor μ is reduced in each in each iteration (e.g., 10) to speed up the convergence.

Example 8.10
Minimize $f(x) = (x)^2$

Subject to $g(x) = x - 1 \geq 0$ using Kuhn–Tucker method with interior point constraint handling approach.

Solution
We develop the following Lagrangian:

$$L = f(x,u) + \lambda h(x,u) + \mu \ln g(x,u)$$

$$L = (x)^2 + \mu \ln(x-1)$$

$$\frac{\partial \pounds}{\partial x} = 0 \quad 2x + \frac{\mu}{(x-1)} = 0 \Rightarrow 2x^2 - 2x + \mu = 0$$

To obtain the optimal value of x, we solve $f(x) = 2x^2 - 2x + \mu = 0$ using the Newton method:

$$\Delta x = -\frac{f}{f'} = -\frac{2x^2 - 2x + \mu}{4x - 2}$$

We start the solution with an initial condition of $x = 0$ and a value for μ like 10 and reduce μ as the solution moves toward convergence.

Starting from $x = 0$ we have

$$\Delta x = -\frac{f}{f'} = \frac{-(10)}{(-2)} = 5$$

$$x = 5$$

We now reduce μ to 1 and continue:

$$\Delta x = -\frac{f}{f'} = -\frac{\left[2(5)^2 - 2(5) + 1\right]}{\left[4(5) - 2\right]} = \left(\frac{41}{18}\right) = -2.27$$

$$x = 5 + (-2.27) = 3.73$$

We will continue with the iterative process, increasing the value of μ in each iteration until the solution convergence is obtained.

Example 8.11

Minimize the following function using the interior point method:

$$f(x) = 6x_1^2 + 3x_2^2$$

subject to

$$g_1(x) = -x_1 + x_2 \geq 0.4$$

Solution

$$c_1(x) = -x_1 + x_2 - 0.4 = 0$$

$$L = f(x) - \zeta \ln(c_1(x))$$

$$L = 6x_1^2 + 3x_2^2 - \zeta \ln(-x_1 + x_2 - 0.4)$$

$$\nabla L = \begin{bmatrix} 12x_1 + \dfrac{\zeta}{-x_1 + x_2 - 0.4} \\[2mm] 6x_2 - \dfrac{\zeta}{-x_1 + x_2 - 0.4} \end{bmatrix} = \begin{bmatrix} 0 \\ 0 \end{bmatrix}$$

The gradient of L is obtained as

$$h(x) = \nabla L = \begin{bmatrix} 0 \\ 0 \end{bmatrix} = \begin{bmatrix} 12x_1(-x_1 + x_2 - 0.4) + \zeta \\ 6x_2(-x_1 + x_2 - 0.4) - \zeta \end{bmatrix}$$

$$h(x) = \begin{bmatrix} 12(-x_1 x_1 + x_2 x_1 - 0.4x_1) + \zeta \\ 6(-x_1 x_2 + x_2 x_2 - 0.4x_2) - \zeta \end{bmatrix}$$

The Jacobian is obtained as

$$h'(x) = \begin{bmatrix} -24x_1 + 12x_2 - 4.8 & 12x_1 \\ -6x_2 & -6x_1 + 12x_2 - 2.4 \end{bmatrix}$$

Solving using these steps each iteration ζ decreases by a factor of 0.1:

$$x_{n+1} = x_n - (h'(x))^{-1} h(x)$$

$$x_{n+1} = x_n - \begin{bmatrix} -24x_1 + 12x_2 - 4.8 & 12x_1 \\ -6x_2 & -6x_1 + 12x_2 - 2.4 \end{bmatrix}^{-1} \begin{bmatrix} 12x_1(-x_1 + x_2 - 0.4) + \zeta \\ 6x_2(-x_1 + x_2 - 0.4) - \zeta \end{bmatrix}$$

We start iterations with a starting value of $x_1 = -0.133$, $x_2 = 0.266$, and $\zeta = 0.01$ and reduce ζ as iterations progress by a factor. Table 8.3 provides the calculation details.

TABLE 8.3 Calculation details of Example 8.11

$J=n$	ζ	x_n	$h'(x)$		$(h'(x))^{-1}$		$h(x)$	$(h'(x))^{-1}h(x)$	x_{n+1}
0	10^{-2}	−0.1330	1.584	−1.596	0.490	−0.140	0.012	0.007	−0.140
		0.2660	−1.596	−5.592	−0.140	−0.139	−0.012	0.000	0.266
1	10^{-3}	−0.1403	1.760	−1.684	0.448	−0.133	−0.010	−0.006	−0.135
		0.2660	−1.596	−5.680	−0.126	−0.139	0.009	0.000	0.266
2	10^{-4}	−0.1348	1.628	−1.617	0.479	−0.138	−0.001	−0.001	−0.134
		0.2661	−1.596	−5.614	−0.136	−0.139	0.001	0.000	0.266
3	10^{-5}	−0.1340	1.609	−1.608	0.484	−0.139	0.000	0.000	−0.134
		0.2661	−1.596	−5.604	−0.138	−0.139	0.000	0.000	0.266
4	10^{-6}	−0.1339	1.607	−1.607	0.484	−0.139	0.000	0.000	−0.134
		0.2661	−1.596	−5.604	−0.138	−0.139	0.000	0.000	0.266
5	10^{-7}	−0.1339	1.607	−1.607	0.484	−0.139	0.000	0.000	−0.134
		0.2661	−1.596	−5.604	−0.138	−0.139	0.000	0.000	0.266

8.4.1.6 OPF Extensions OPF methods need to be extended to be applied to different types of solutions. Some of these extensions include the following:

1. **Multitime points**: Since OPF provides the solution for a single time point, it needs to be extended to deal with applications that spans over a certain period of time [64] such as a unit commitment that will need to take into account time performance of different units as they are committed or decommitted (e.g., start time, ramping rate).

2. **Discrete decisions**: In some problems, the decision would need to be made that involves zero or one decisions. Special optimization extensions are needed to deal with these problems. Examples of these problems are control discretization for switchable shunts [65, 66, 67] and transformer taps as well as unit commitment dealing with introducing or terminating generator units.

3. **Multinetwork**: These problems involve the solution of a number of networks together. An example of this solution is security constrained OPF [58,68] in which the solutions of the normal system and a number of contingencies need to be solved together to provide a feasible solution.

Figure 8.9 summarizes OPF classifications and extensions discussed earlier.

8.4.2 Generation Sufficiency Infrastructure

The infrastructure depends on the utility model as given next:

- In the vertically integrated utilities, the generation solutions could be co-located in the control center with no barrier between transmission and generation information.

- In the restructured vertically integrated utilities, the generation solution infrastructure is located in the generation division, and there is an arm's-length relationship between generation and the transmission.

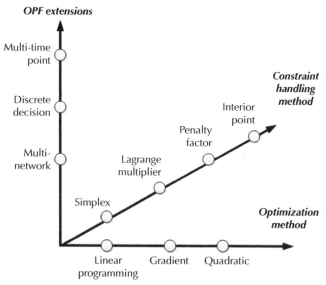

Figure 8.9 OPF classifications and extensions.

- In restructured utilities, the market solutions reside on the market operation system, which may or may not be located at the control center, depending on whether the system operation administers the market operation.

QUESTIONS AND PROBLEMS

8.1. List four dominant features of the restructured systems.

8.2. Draw the functional entities in a restructured vertically integrated utility.

8.3. What is one reason for the existence of the restructured utility model?

8.4. What are regulating, fast, and slower reserves?

8.5. List three optimal power flow extensions.

8.6. Minimize the following function using the simplex method

$$f(x) = x_1 - 2x_2 - x_3$$

subject to

$$2x_1 + x_2 - x_3 \geq 2$$
$$2x_1 - x_2 + 5x_2 \geq 6$$
$$4x_1 + 1.5x_2 + 1.5x_3 \geq 6$$

8.7. Minimize the following function using the gradient method for one iteration starting at (1, −2/3):

$$f(x) = \frac{1}{3}x^3 - x$$

8.8. Minimize the following function using Newton's method for one iteration starting at $(1, 4/3-)$:

$$f(x) = \frac{1}{3}x^3 + x$$

8.9. Minimize the following function using the Kuhn–Tucker method

$$f(x) = x_1^2 + x_2^2$$

subject to

$$g_1(x) = -x_1 + x_2 \le 0.4$$

8.10. Minimize the following function using the interior point method:

$$f(x) = x_1^2 + x_2^2$$

subject to

$$g_1(x) = -x_1 + x_2 \ge 0.4$$

8.11. Minimize the following function using the penalty method

$$f(x) = x_1^2 + x_2^2$$

subject to

$$g_1(x) = -x_1 + x_2 \ge 0.4$$

CHAPTER 9

SYSTEM OPERATION CONTROL CENTERS

9.1 INTRODUCTION

Control centers are command and control locations where system operation takes place. Critical processes associated with transmission operation, distribution operation, market operation, and restoration are executed at the control center. Some of these critical processes may have a service restoration requirement of 30 minutes and yearly availability of 99.95%. This would mean that the technology supporting the process cannot be out of service by more than 230 minutes or 3.83 hours in 1 year. Thus, the control center infrastructure should be designed to support the availability requirement of the technology solution.

The high availability requirement poses a number of specific design attributes on the control center infrastructure design to ensure that it can withstand events such as natural disasters, accidents, environmental and biological incidents, as well as malicious attacks.

Modern control center designs are different from the older ones in two ways:

1. Older control centers, such as the one shown in Figure 9.1, did not fully subscribe to many design attributes of the modern control centers.

2. Older control centers had more limited functionality with less region of influence resulting different control center for different regions.

This chapter first describes the criteria and required attributes of modern control centers. It then focuses on the modern control center configurations providing the required redundancy. Finally, it describes the modern control center configuration and design trend that satisfies the required attributes.

9.2 MODERN CONTROL CENTER ATTRIBUTES

The most fundamental attribute for a modern control center design is the service availability measured by service restoration time and yearly availability. For example, BC Hydro's modern control center design was based on a service restoration time of

Practical Power System Operation, First Edition. Ebrahim Vaahedi.
© 2014 The Institute of Electrical and Electronics Engineers, Inc. Published 2014 by John Wiley & Sons, Inc.

Figure 9.1   BC Hydro's old control center.

15 minutes and a yearly availability of 99.95%. This performance requirement constitutes two important facts:

1. Operators could not tolerate losing the monitoring and control functionality of the control center for more than 15 minutes.

2. Operators could not tolerate suffering more than 15 outages per year resulting in 99.95% availability.

A typical list of the events that the control center needs to be designed for is given in Table 9.1. This table shows the event risk along the design requirement to mitigating the risk associated with the event. It should be noted that a proper method should be used to assess the risk for different events, which can be calculated by the probability of the event occurrence multiplied by the impact of the event. If the risk associated with an event is negligible or low, then that event does not need to be considered in the control center design and vice versa. The outcome of Table 9.1 determines the control center design and redundancy requirements.

As an example, if the risk of earthquake for a utility is not negligible, the utility needs to mitigate its risk by considering the following mitigating design features:

1. The building infrastructure needs to be built to a higher standard of resiliency against earthquakes.

2. A backup control center needs to be developed in a different seismic region to mitigate the risk of severe earthquakes that the infrastructure has not been designed for.

3. The development of a disaster recovery plan to recover from an earthquake.

TABLE 9.1   Events considered for control center design and potential mitigation solutions

Event type	Event	Event risk	Required design to mitigate event risk
Natural	Earthquake		Resilient infrastructure, a secondary control center in a different seismic area, disaster recovery procedures
	Fire		Fire suppression equipment, a secondary control center, disaster recovery procedures
	Flood		A secondary control center, disaster recovery procedures
	Storms		A secondary control center, disaster recovery procedures
	Electromagnetic storms		A secondary control center, disaster recovery procedures
Accidental	Car accident		Concrete barriers around the building, disaster recovery procedures
	Aircraft incident		Resilient construction, a secondary control center, disaster recovery procedures
	Explosion		Resilient construction, shatter proof glasses and steel reinforced windows, a secondary control center, disaster recovery procedures
	Human error		Training facilities, disaster recovery procedures, a secondary control center
Malicious attack	Terrorist attack		Physical security barriers, entrance gate, concrete barriers, 24/7 security, double authentication, compliance with NERC physical security standards, disaster recovery procedures
	Cyber security attack		Security architecture designs, compliance with NERC cyber security standards, disaster recovery procedures
	Physical security attack		Physical security barriers, entrance gate, concrete barriers, 24/7 security, double authentication, a secondary control center, disaster recovery procedures
	Vandalism		Compliance with NERC cyber and physical security standards, disaster recovery procedures
Biological or environmental	Disease		Safety procedures, a secondary control center, disaster recovery procedures
	Air contamination		Safety procedures, a secondary control center, disaster recovery procedures
	Water contamination		Safety procedures, a secondary control center, disaster recovery procedures
	Toxic spill		Safety procedures, a secondary control center, disaster recovery procedures

9.3 CONTROL CENTER REDUNDANCY CONFIGURATION

Redundancy at the control center is achieved through two different architectures:

1. ***Duplicate systems at the same site***: This architecture allows the operation of two similar systems in one control center with one system operating on the standby mode ready to take over the operation automatically if the main operating system fails.

2. ***Duplicate control centers***: This architecture promotes the operation of two control centers at two different sites. It come in two different flavors:

   a. Two hot sites with one acting on standby and ready to automatically take over the operation in a few minutes if the main site fails.

   b. One hot site and one cold standby taking over the operation in a few hours once the main site fails.

Due to its long restoration time, the cold site may not fulfill the redundancy requirement.

Figure 9.2 shows the redundancy architecture described earlier existing within one control center site and between two control centers. Systems A, B, C, and D are exactly the same systems. If system A fails, system B will take over. On the other hand, if the whole primary control center goes out due to some event such as an earthquake, then the second site with systems C and D will take over.

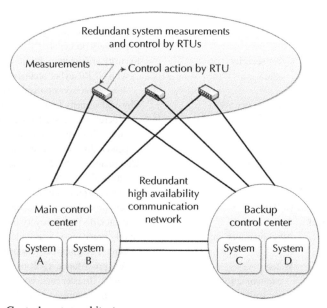

Figure 9.2   Control center architecture.

9.4   MODERN CONTROL CENTER CONFIGURATION

In the past, control centers had a limited focus. It was not unusual for a utility to have many control centers for transmission operation and regional operations [69]. With the advancement in technology, the trend for modern control center design has shifted to the consolidation of functions of transmission and distribution control centers [69]. The trend is to reduce the number of control centers to a minimum with the following justifications:

- Reducing control center building costs
- Reducing computer system costs
- Reducing operating and maintenance staff costs
- Achieving better coordination of operations across the power system
- Achieving uniformity of operation for transmission and distribution

These justifications along with the requirement to mitigate different event risks typically advocate a two-control center configuration shown in Figure 9.2. This comes in two different flavors:

1. Two fully staffed control centers sharing different tasks on a daily basis.
2. The backup control center is partially staffed, requiring staff augmentation when the backup control center takes over the operation from the primary one.

The second solution provides efficiencies in terms of staff headcount. The location of the backup control center also needs a number of considerations:

1. The backup control center should be located in a different seismic region than the primary control center to reduce the risk of both control centers being taken down by the same earthquake.
2. The backup control center should be in a location that has high availability communication system for the backup control center monitoring and control functions as shown in Figure 9.2. For example, BC Hydro's backup control center location was selected to be on BC Hydro's owned highly reliable communication ring system going around the province.
3. If the backup control center is not fully staffed, then the staff augmentation time during disasters may restrict the distance.

Figure 9.3 shows the inside of BC Hydro's modern control center, which manages transmission operation, distribution operation, and generation operation.

The inside layout of a control center is generally arranged by job function. For example BC Hydro's control center is configured in the following layout:

- **Distribution desks** are occupied by load operators who manage distribution circuits. When the power goes out, load operators are responsible for directing activities to get it back on again. Because there is always maintenance and capital work taking place in substations and on the distribution system, they are also responsible for issuing safety permits that allow the work to be done safely by the field staff.

Figure 9.3   BC Hydro's modern control center.

- **Transmission desks** occupied by grid operators and transmission coordinators monitor high-voltage transmission lines and substations. Large amounts of power flow on these lines at extremely high voltages. Grid operators need to make sure that the lines are loaded within their limits as well as be prepared for what would happen if any one of these lines suddenly became de-energized. Transmission coordinators are responsible for coordinating transmission outages.

- **Interchange, plant operations, and generation coordination desks** are occupied by interchange operators, plant operators, and generation coordinators. The interchange operator manages the flow of energy in and out of the province, while the plant operator controls most of the generating plants and monitors the water levels in our dams and along our rivers. The generation coordinator controls the output of the very large plants and makes sure they are responding to minute-by-minute changes in demand.

- **System control desk** is occupied by system control managers who supervise all the roles within the control center.

9.5   MODERN CONTROL CENTER DESIGN DETAILS

The salient features associated with the modern control center design include

- **Physical security**: The modern control center should be able to protect the critical systems. The important features associated with physical security include
  - High barriers (e.g., 12 ft) around the perimeter
  - Low concrete barriers (e.g., 3 ft) around the perimeter
  - Entrance gate

Figure 9.4   One of the two diesel generators at BC Hydro's control center.

- ○ 24/7 security guards
- ○ Double authentication locks, biometric and electronic cards
- ○ Earthquake-resilient buildings (e.g., 1 in 10,000 year earthquake)
- ○ Earthquake-resistant computer racks
- ○ Shatter-proof window glasses
- ○ Steel-reinforced windows
- **Ancillary services**: Each control center should have redundant ancillary systems to support continued availability of operations:
  - ○ Redundant diesel generators with sufficient fuel for up to 2 weeks to support the control center if the main electricity supply to the control center fails. Figure 9.4 shows one of the two 1.25 MW diesel generators available at each of BC Hydro's modern control centers.
  - ○ Battery sets to keep the operation for 1.5–4 hours until a diesel generator can be started up as shown in Figure 9.5.
  - ○ Redundant air handling units.
  - ○ Redundant chillers.
- **Control room features**: The control room needs to be designed with features to maximize the operator's comfort, sometimes working in long shifts:
  - ○ Height-adjustable consoles allowing the operator to work in a sitting or standing position as shown in Figure 9.6. The console is equipped with personal air-conditioning units.

Figure 9.5   BC Hydro's battery room.

Figure 9.6   A typical console at BC Hydro's control center.

Figure 9.7   IP addressable light and noise reduction wall at BC Hydro's modern control center.

- ○ A number of computer terminals providing a full coverage of the system transmission diagram acting as a personal wall diagram for each operator.
- ○ Internet protocol (IP) addressable lights to provide light intensity and light color choice for each operator as shown in Figure 9.7.
- ○ Noise reduction walls as shown in Figure 9.7.
- • **Compliance requirement**: Due to the nature of the critical processes executed at the control, NERC has established physical and cyber standards that need to be adhered to.

QUESTIONS AND PROBLEMS

9.1. What is the typical performance requirement for a control center?

9.2. What are the four broad categories of events to consider for the control center design?

9.3. Consulting Table 9.1, describe the design features needed to mitigate the impact of storms and diseases?

9.4. What are the two approaches to provide redundancy?

9.5. Why do we need control center redundancy?

9.6. What is the configuration trend in modern control centers?

9.7. What is the difference between hot redundancy and cold redundancy?

9.8. What are the objectives for consolidation of control centers?

9.9. What is the redundancy configuration of a modern control center?

9.10. What are the considerations for selecting the location of the backup control center?

9.11. What is the need for batteries at the control center?

9.12. How much fuel should the diesel generators at a control center have?

CHAPTER 10

ENERGY MANAGEMENT SYSTEMS

10.1 INTRODUCTION

Energy management systems (EMS) are real-time computer systems that were initially introduced in the early 1970s [70–72, 75] to provide system operators with the means to manage the power system grid in a reliable and efficient manner. Energy management systems achieve their objectives by providing decision support systems and control means for generation and transmission systems. Similar solutions have been devised for distribution systems called distribution management systems (DMS), which will be discussed in Chapter 11.

In the initial decades following their introductions, the hardware and software requirements presented major challenges to the vendor to have acceptable system performance requirements. This led to vendors choosing very customized and proprietary solutions that were difficult to sustain both by customers and vendors as technology solutions advanced. Utilities were left with an enormous total cost of ownership (TCO), which reflects both the initial costs and the cost of sustainment during the product's useful life. This in turn led to a number of vendors going out of business or convergence of vendors and utilities abandoning or overhauling their older energy management systems.

With advances in technology in the 1990s, this situation started to change. Vendors took advantage of the advancements to develop baseline EMS solutions that can be deployed on many customers without too much customization, thereby reducing the cost of the EMS. Software advancements have also resulted in advancements in the user interface and decision support modules, making the energy management systems of today much richer than their initial offerings. As an example, the utility restructuring resulted in the creation of the market management systems, which can be considered a major decision support tool of the energy management systems.

Due to the critical tasks performed by the EMS, these systems have a strict system availability requirement. For this reason, a lot of attention is placed in choosing the appropriate architectures to achieve the availability requirement.

Figure 10.1 shows BC Hydro's energy management system providing many display monitors to operators for increased situational awareness and decision support.

Practical Power System Operation, First Edition. Ebrahim Vaahedi.
© 2014 The Institute of Electrical and Electronics Engineers, Inc. Published 2014 by John Wiley & Sons, Inc.

Figure 10.1   BC Hydro's Energy Management System. Published with permission from Charles Woolfries.

10.2   EMS FUNCTIONALITY OVERVIEW

As shown in Figure 10.2, the EMS functionality can be divided into three categories:

- System monitoring
- Decision support tools
- Control

Furthermore, this functionality is offered in the following domains:

- Real-time production system
- Operations planning production system
- Study systems
- Training simulator system
- Quality assurance system

A real-time production system continuously operates to provide decision support and control solutions for managing the real-time system operation. Similarly, an operations planning production system provides decision support and control solutions for managing the operations planning processes.

Study systems are put in place to perform what-if studies. For example, operations planning engineers may examine alternatives other than those produced by the production systems to establish whether they can safely take a transmission line out.

The training simulation domain deals with an environment that caters to providing scenarios for training system operators. These scenarios are executed using scripts to provide system conditions that operators face and the actions they need to take.

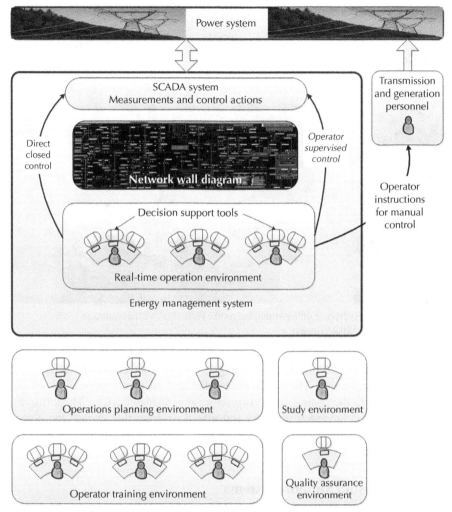

Figure 10.2   EMS functionality and functional environments.

Finally, the quality assurance system is used for testing new applications and upgrades before introducing them to the production system (Fig. 10.3).

10.2.1   System Monitoring

This functionality provides an accurate picture of the system conditions. As described in Chapter 2, the SCADA portion of the EMS obtains information regarding the current operating condition of the power system by continually gathering data at key locations. These data such as MW values or circuit breaker positions are transmitted to the system control center over a telemetry system. The acquisition and teleme-tering activities of the SCADA are performed cyclically every few seconds, which enables control supervisors to have as much instantaneous views of the system as

Figure 10.3   BC Hydro's training simulator room. Published with permission from Brett Hallborg and Mike Gimson.

possible. The EMS performs data processing activities upon receipt of power systems data at the control center. The power system data and the results of the analysis are made available to the control supervisor on various display devices. These display devices include color video monitors and a wall diagram that depicts the power system network in simplified form. The data on these display devices are updated automatically in real time.

10.2.2   Decision Support Systems

This functionality provides a rich set of solutions to ensure system reliability and efficiency [74]. These tools generally use power system information and system models to provide security and efficiency solutions, as well as to provide appropriate control actions. Chapters 2 to 8 elaborated on a number of these tools. A more complete list reflecting a typical set of decision support tools in an EMS will be described later in Section 10.3.2.

10.2.3   EMS Control Actions

There are three types of control actions implemented by the EMS as shown in Figure 10.1:

1. *Direct closed control*: This control action is implemented directly by EMS with no operator intervention. An example of this kind of control action is the automatic generation control, which is implemented automatically based

on frequency and interchange schedule deviations. Another example of this control is the automatic remedial action schemes, which are designed to take place following specific system conditions such as the loss of a major transmission line.

2. *Operator supervised controls*: These controls are implemented through EMS by operators. An example of this control is an operator opening a SCADA controlled switch.

3. *Operator manual control*: These controls are implemented manually by field staffs after receiving orders from the operators. An example of this control is the change of generation at a plant recommended by the EMS. The operator will need to call a plant operator to change the generation because he or she does not have direct control of the plant.

10.3 ENERGY MANAGEMENT SYSTEM AVAILABILITY CRITERIA AND ARCHITECTURE

Transmission operation represents one of the most critical processes in utilities. EMS enables the transmission operation process by providing operators with system visibility and control. Without the EMS functionality, operators can manage transmission operation for about 30 minutes. The main reason for the 30 minute time frame stems from the fact that system load can change substantively within 30 minutes. A substantive load change makes the predetermined remedial actions ineffective. Furthermore, not having the visibility and control removes the possibility of the operator manually steering the system. For this reason, if the EMS is unavailable beyond 30 minutes, operators will have to resort to operating the system in a more conservative manner.

Since the EMS is a mission-critical system, there are three different criteria used to ensure its availability and business continuity for prolonged outages:

1. **Restoration time**: The first criterion for system availability is restoration time following an EMS outage. The energy management system restoration time is normally selected to be around 30 minutes.

2. **System availability**: The second criterion represents the system availability percentage during the year. A system availability of 99.95% represents a total of 263 minutes during the year. The first criterion for system availability is restoration time following an EMS outage. The energy management system availability is normally selected to be around 99.95%.

3. **Business continuity**: For those unlikely situations where the system may have prolonged EMS outages, business continuity processes and solutions need to be devised. The business continuity process ensures that the operators continue to manage the system operation in a more conservative way to manage information unavailability.

The requirement of the EMS to provide high availability for critical functions forces a double redundant design or a quadruple design shown in Figure 10.4. This figure shows

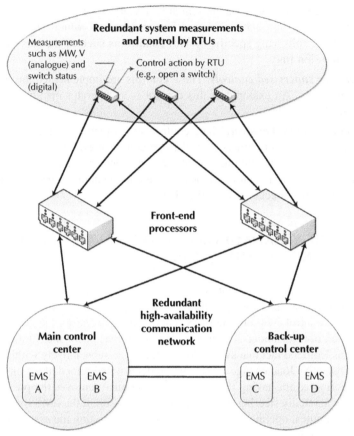

Figure 10.4 EMS architecture.

how information flows from remote terminal units (RTUs) to the control centers. Normally, front end processors are placed at a few locations in the system to collect information from RTUs and send them over to the control centers to avoid too many direct connections from RTUs. This figure indicates that we have redundant hardware and software at the main control center and the backup control center. If there is a hardware failure at either location, the redundant hardware will take over the operation. If there is a major emergency at the main control center making it nonoperational, then the system operation will be failed over to the backup control center. The communication systems connecting the RTUs to the control centers and those connecting the two control centers together are designed to have high availability and redundancy.

10.3.1 Hardware Overview

EMS systems comprise the following four subsystems:

- Data acquisition subsystem
- Computer subsystem

- User subsystem
- Auxiliary power supply

These subsystems are described next.

10.3.1.1   *Data Acquisition Subsystem*   The task of collecting and transmitting data to the control center is performed by the data acquisition system as described in Chapter 2. Major power system parameters such as active and reactive power, voltage, frequency, water levels, and circuit breaker positions are transmitted to the control center to update the database model of the power system. Two main components of this system are the RTUs and the communication network subsystem (CNS).

The actual collection of data is performed by RTUs. The field instrumentations, such as transducers or pallet switches in breakers, are wired to terminal racks, which in turn are connected to RTUs at stations. The analog signals from transducers (e.g., quantitative values such as MW) are presented to the RTU as values between −5, 0, +5 DC and are converted to digital values by the analog to digital (A/D) converter within the RTU. Status points such as circuit breaker positions are input to RTU as shown in Figure 10.4. RTUs also implement control actions sent by the control center.

This subsystem is responsible for the transmission of data from the RTUs to the control center and the transmission of control action to be taken by RTUs from the control center. It is designed such that no single contingency faults will result in the loss of communication with an RTU.

10.3.1.2   *Computer Subsystem*   The computer subsystem provides the following functions:

1. SCADA: data acquisition, alarm processing, and user interface (updating of displayed information on computer monitors and wall diagrams).
2. Database: storage and retrieval services.
3. Processing of application programs.

The main requirements of the computer subsystem are

- The total processing requirement is much higher for network analysis than for all other parts combined.
- All areas require a high access rate of the database.
- The SCADA function must respond to a high interrupt load from external system.

These diverse requirements call for a special hardware design. Computer performance and computer interrupt processing are two different features not easily accomplished within a single computer.

SCADA load characteristically consists of a large number of small and quick tasks. The growth in SCADA takes the form of horizontal growth replicating components to add more monitors and RTUs, while growth in network analysis takes the form of vertical growth expanding the size of the computer used for network analysis. To accommodate these growths, the SCADA portion should be expanded incrementally as new devices are added, and the network analysis computers should be upgraded or replaced to accommodate additional functions and bigger power system models. Because of these different requirements, different architectures have appeared [75]. The differences in these architectures lie in mixing different computers to take care of performance and interrupt capacity.

10.3.1.3 User Interface Subsystem The information collected and processed by the EMS is made available to control supervisor via man/machine subsystem. The MMS as shown in Figure 10.1 mainly consists of several consoles and a wall diagram.

10.3.1.4 Auxiliary Power Subsystem This system, which is also called uninterruptible power supply, provides the control center electric power continuity during the times that the main electricity supply to the control center is disrupted. The auxiliary power subsystem is composed of two elements:

1. A number of diesel generator units with sufficient fuel to be able to supply the control center electricity by up to about 15 days.

2. A number of batteries to provide electricity for the duration of between 1 and 4 hours to provide the diesel generator enough time to start up.

10.3.2 Software Overview

The EMS software consists of various decision support tools as well as other administration and support functions needed for EMS to work. Figure 10.5 provides a typical EMS software application list. A short description for each application is provided next:

1. *SCADA*: Supervisory control and data acquisition system performs data acquisition, alarm processing, user interface updating, as well as execution of control actions in the field.

2. *Network configuration*: This tool is responsible for maintaining a single-line diagram model of the bulk electricity supply.

3. *State estimation*: This is a mathematical method that uses available power system measurement values to recreate values for all other unknown system state variables.

4. *Remedial action system (RAS)*: These programs are intended to assist the operators in arriving at appropriate remedial control actions to correct for any security violation in the normal system condition and after credible contingencies.

5. *Scenario analysis (power flow)*: This uses the system model, generation, and load values to calculate power flows and voltage values at all other buses in the system.

6. *Static security (contingency analysis)*: This is a method to identify the system's thermal and voltage violations during normal conditions and after credible contingencies.

7. *Transient stability*: This tool provides transient stability limits and remedial action recommendations for operators.

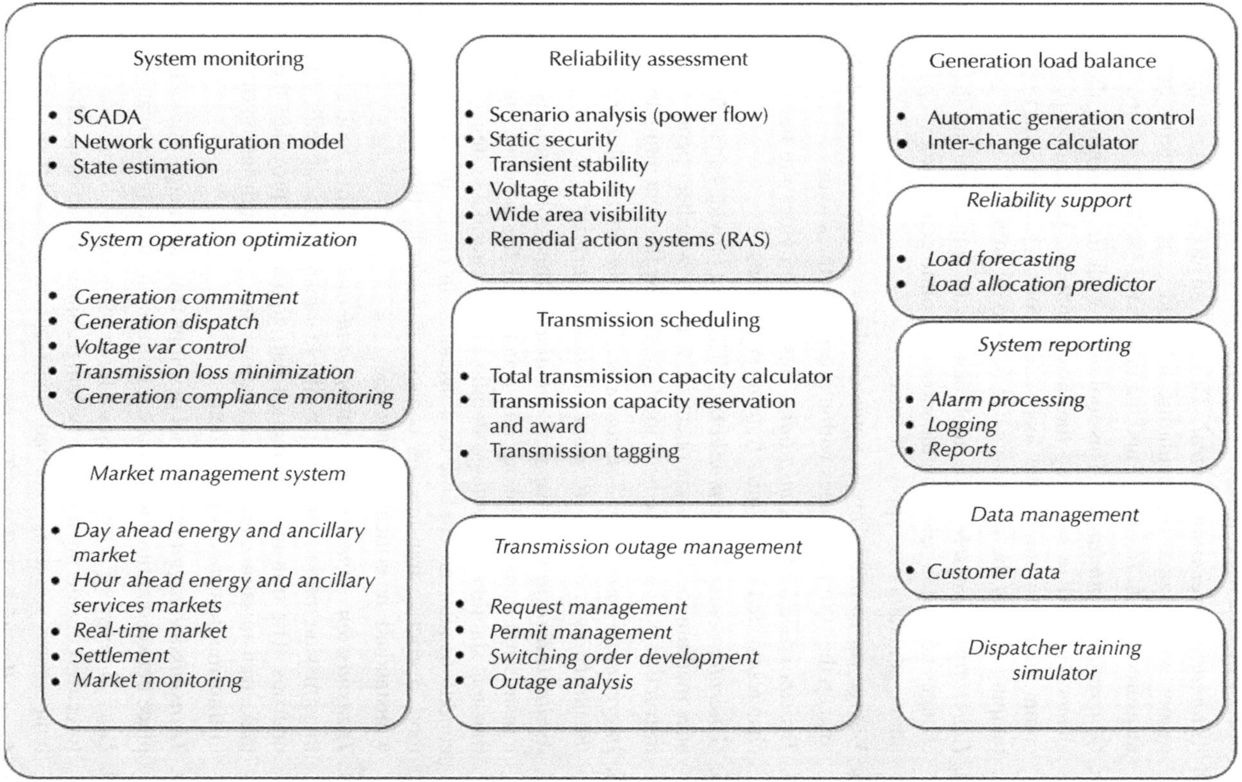

Figure 10.5 A typical EMS application portfolio.

System monitoring
- SCADA
- Network configuration model
- State estimation

System operation optimization
- Generation commitment
- Generation dispatch
- Voltage var control
- Transmission loss minimization
- Generation compliance monitoring

Market management system
- Day ahead energy and ancillary market
- Hour ahead energy and ancillary services markets
- Real-time market
- Settlement
- Market monitoring

Reliability assessment
- Scenario analysis (power flow)
- Static security
- Transient stability
- Voltage stability
- Wide area visibility
- Remedial action systems (RAS)

Transmission scheduling
- Total transmission capacity calculator
- Transmission capacity reservation and award
- Transmission tagging

Transmission outage management
- Request management
- Permit management
- Switching order development
- Outage analysis

Generation load balance
- Automatic generation control
- Inter-change calculator

Reliability support
- Load forecasting
- Load allocation predictor

System reporting
- Alarm processing
- Logging
- Reports

Data management
- Customer data

Dispatcher training simulator

8. *Voltage stability*: This tool provides voltage stability limits and remedial action recommendations for operators.

9. *Wide-area visibility*: This tool provides situational awareness to system operators on a wide area.

10. *Automatic generation control*: This tool ensures minute-to-minute generation load balance by regulating selected generating units based on the measured system frequency and the interchange schedule errors.

11. *Generation commitment*: This tool deals with implementing the necessary changes needed in selecting the most economic generation portfolio (unit commitment) to supply the load while considering the start-up time and ramping rates of the generators that need to go online.

12. *Generation dispatch*: This tool deals with implementing the necessary changes needed in selecting the most economic generation set points for the online generators to supply the load.

13. *Voltage var control*: This tool provides the necessary control changes needed to keep the system voltages within acceptable thresholds.

14. *Transmission loss minimization*: This tool provides the necessary control changes needed to minimize transmission losses.

15. *Generation compliance monitoring*: This tool provides an audit and compliance medium for the system operators to check whether operators' generation instruction changes have been implemented by the generating units.

16. *Interchange calculator*: *This tool calculates the interchange transactions on the utility interties for settlement and billing purposes.*

17. *Available transmission capacity calculator*: This tool calculates the available transmission capacity taking into account all the security limits (thermal, transient stability, and voltage stability) for the normal system conditions and conditions after all credible contingencies. In effect, the tool calculates the most restrictive transmission capacity considering that the system should withstand all the critical contingencies.

18. *Transmission capacity reservation and award*: This tool administers a transmission market reflecting available transmission capacities on different interties. The transmission market participants are provided the ability to make transmission capacity reservations and get confirmation of the amount of transmission capacity awarded.

19. *Transmission tagging*: This tool develops tags for energy transactions taking place between entities to specific transmission capacity reservations.

20. *Load forecasting*: This tool uses a forecasting model to provide system load forecasts considering different system parameters such as weather conditions, temperature, and temporal conditions (season, day, and hour).

21. *Load allocation predictor*: This tool allocates the substation load to different feeders using historical models.

22. *Day ahead energy and ancillary services market*: This tool, which is part of the market management system, provides day ahead generation schedules to

the operator by administering the day ahead energy and ancillary services markets for the market participants.

23. *Hour ahead energy and ancillary services market*: This tool, which is part of the market management system, provides hour ahead generation schedules to the operator by administering the hour ahead energy and ancillary services markets for the market participants.

24. *Real-time energy and ancillary services market*: This tool, which is part of the market management system, provides real-time generation schedules to the operator by administering the real-time energy and ancillary services markets for the market participants.

25. *Settlement system*: This tool calculates the financial settlement among different parties.

26. *Market monitoring*: This tool monitors market information and establishes market design and behavior anomalies by participants.

27. *Transmission outage management*: This tool provides a collaborative tool between the control center and the field staff responsible for implementing outages. The tool provides the user the ability to submit transmission outage requests, analyze outage impacts, approve outages, and develop switching orders for implementing the outage.

28. *System reporting*: These tools collect data to produce alarms and necessary online and historical data summary displays and reports.

29. *Data management*: This tool manages customer contact information and other operational data.

30. *Dispatcher training simulator*: This tool provides a realistic environment for hands-on dispatcher training under simulated normal, emergency, and restorative operating conditions. The training is based on interactive communication between instructor and trainee with a complete replica of the DMS user interface. The training simulator has its own separate environment.

10.3.3 Application Sequencing in EMS

The application programs in EMS are normally placed in a cyclic sequence to execute a complete process. As shown in Figure 10.6, once an application in the sequence is executed, a time stamp is generated. The sequence time cycle depends on the process that the sequence is designed to execute, and it is ensured that the time cycle is chosen in a way that all the applications can be executed within the selected time cycle. This cyclic sequence process continues until a major change happens in the system such as a sudden load change, generation change, or network change resulting from a contingency. Under this condition, the sequence restarts from the beginning considering the prevailing system operating conditions. For example, to execute the process of posturing for static security requires executing state estimation and contingency analysis. Figure 10.7 shows BC Hydro's EMS application sequencing display.

Since different transmission operation processes use different application sequences, it is possible to architect different application sequences sharing some

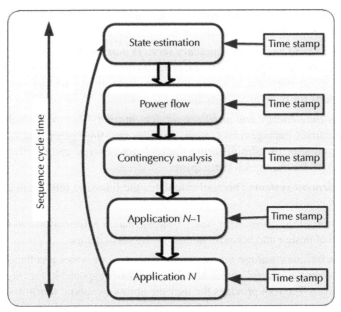

Figure 10.6   An application sequence in EMS.

part of the sequence. As an example, Figure 10.8 shows two application sequences, one branching out from the other, resulting in different cycle times.

10.3.4   Software Integration

Energy management system is one of the major software blocks of a control center. There are other software solutions. To minimize the total cost of ownership, a solution needs to be fit for the present requirement and it should also stand the test of time [69,71]. To ensure that a solution stands the test of time, the maintenance and sustainment costs for changes during the life span of the solution needs to be minimized. For that reason, new architecture solutions based on service-oriented architectures (SOA) have been developed, enabling change in one component without changing the whole solution [69].

A recent CIGRE task force [76] has produced a roadmap for future EMS designs that are open and scalable, enabling users to select the best of the breed from different vendors with seamless integration. The following five salient features constitute the core elements of the new design:

- *Service-oriented architectures (SOA)*: The deployment of SOA constitutes modular and reusable components. This model facilitates integration of third-party software. Furthermore, the decoupled feature of the architecture minimizes the replacement cost.

- *Common information model compatibility*: EPRI's common information model's adoptions by vendors ensure that consistent information models are used and that it would be easy to mix different software products.

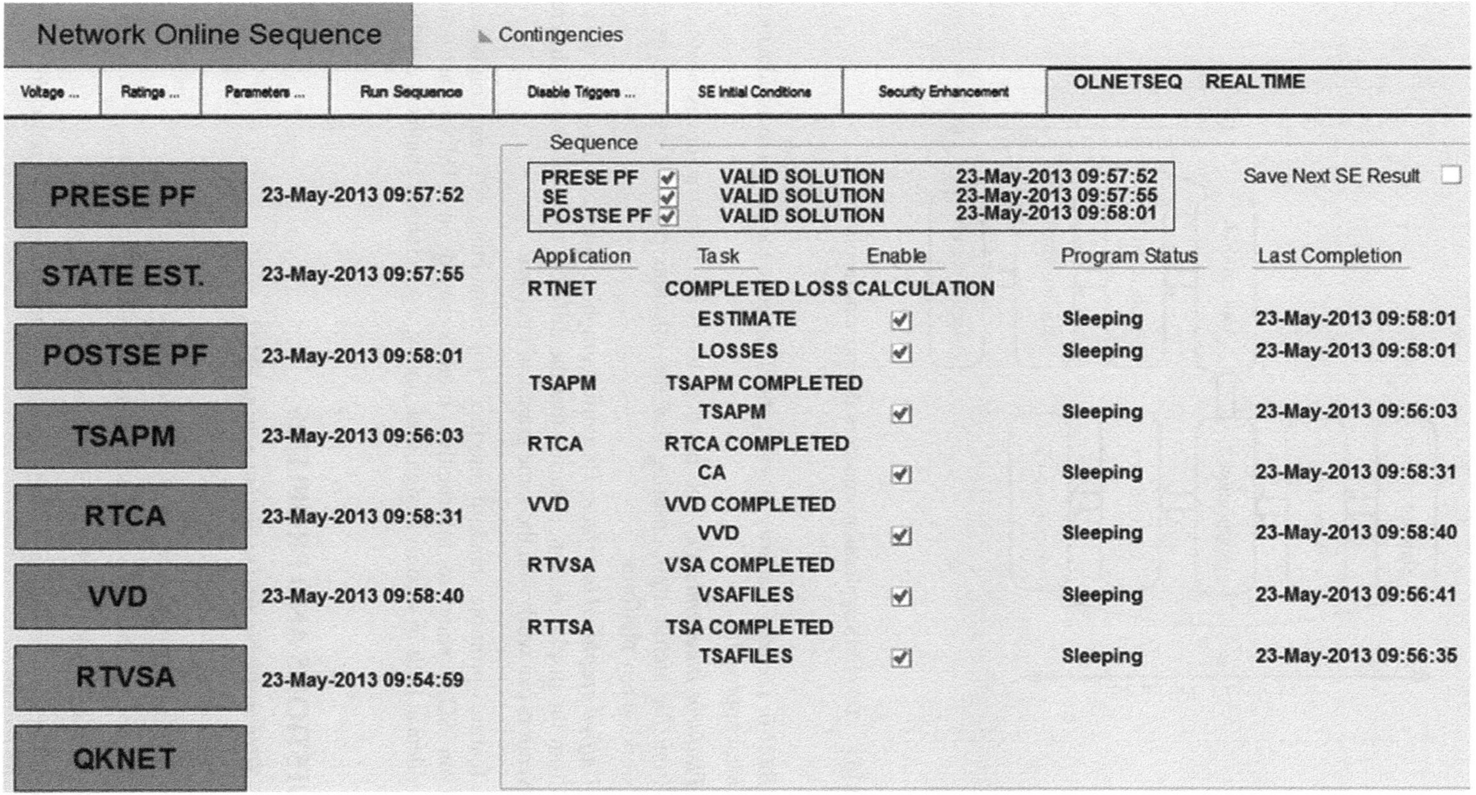

Figure 10.7 BC Hydro's application sequence display.

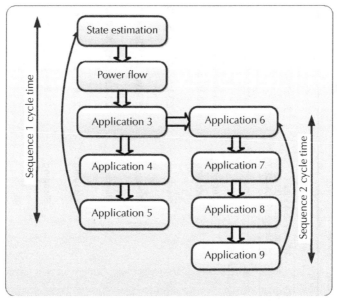

Figure 10.8 Two EMS application sequences sharing some applications.

- **Built-in security**: Security plays a central theme in the design to avoid a piece-meal approach.
- **Platform independence**: The physical hardware infrastructure is abstracted from the architecture, ensuring that the solutions provided by vendors can work on any platform.
- **Unified graphical interface**: The unified graphical interface decouples the tool from the display design, ensuring that the different tools offered by different vendors can work with the same graphical interface.

These features promote standard interfaces for connecting different components based on SOA, standard information models based on CIM, infrastructure independence, built-in security design, and, finally, a generalized graphical interface.

QUESTIONS AND PROBLEMS

10.1. What are the business justifications for an EMS?

10.2. What is the total cost of ownership?

10.3. What are the main categories of EMS functionality?

10.4. Describe the three types of control actions in an EMS.

10.5. What are the domains that the EMS functionality is offered in?

10.6. Being a mission-critical system, what three criteria are used to reflect its required performance?

10.7. Describe the redundant EMS architecture (Figure 10.4).

10.8. Describe the main categories of EMS hardware.

10.9. What is application sequencing in EMS?

10.10. What are the main futures of the proposed CIGRE Task Force Architecture?

10.11. Describe the general categories of software portfolio on an EMS (Figure 10.5).

CHAPTER 11

DISTRIBUTION MANAGEMENT SYSTEM

11.1 INTRODUCTION

Unlike the energy management systems managing the transmission operation, the systems managing distribution operations have evolved very slowly since the 1970s. Up until recent years, almost all the distribution systems were managed using a manual wall board mimic similar to the one shown in Figure 11.1. This wall board contains a static network diagram of the distribution system similar to paper maps and diagrams. Operators manually place sticky notes and comment pages on the wall diagram to reflect prevailing system changes, outages, and other information needed to manage the system. For example, operators need to know on which feeders they have provided permits for system maintenance or which automated equipment should be deactivated to ensure safety of the maintenance field crew. In other words, up until recently, the tasks of data collection, system visibility, and control have been achieved using manual processes.

Not having access to real-time information, system visibility, and control, operators resort to inefficient ways of fulfilling a number of distribution operation processes and functions such as

1. Proper loading of distribution feeders and assessment of spare capacities
2. Assessing the impact of planned outage
3. Developing switching actions a field crew should implement to isolate a line or return it back to service
4. Manually setting system voltages not being able to capture minimal distribution system losses
5. Establishing fault location, fault isolation, and service restoration requiring a great deal of manual involvement by crews

The objectives for distribution management system (DMS) implementation are

- Enhancing safety by providing better visibility and control on system energization and de-energization.
- Extending the useful life span of power system assets by properly managing their operation.

Practical Power System Operation, First Edition. Ebrahim Vaahedi.

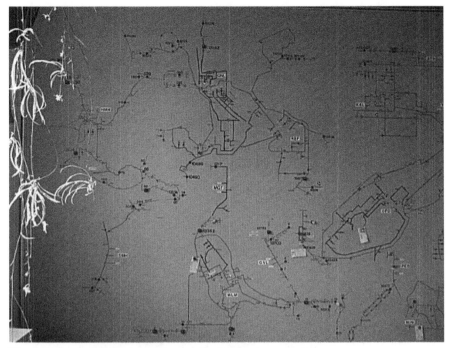

Figure 11.1   A traditional wall diagram for distribution operation.

- Improving system reliability by reducing system outage times.
- Enhancing system efficiency and optimizing the use of available resources.

11.2   DMS FUNCTIONALITY OVERVIEW

DMS is an integrated decision support system whereby all operational aspects of the utility's distribution system are made visible and operable from a central source [77]. Advanced algorithms are used to optimize the system in real time. DMS is the distribution equivalent of the energy management system (EMS), which is used to manage the operations of a transmission system. DMS functionality, as shown in Figure 11.2, can be broadly divided into three categories:

- System monitoring
- Decision support tools
- Control

One of the major differences of DMS functionality with the EMS is that the number of equipment in the distribution system is much more than that of transmission system as elaborated later.

The DMS environments are offered in the following domains:

- Distribution operation environment
- Engineering study environment

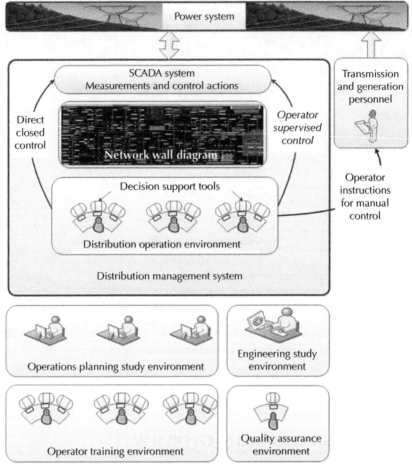

Figure 11.2 DMS functionality and functional environments.

- Operations planning environment
- Training simulator system
- Quality assurance system

The distribution operation environment provides operators with system visibility, decision support, and control for managing the distribution operation. The operations planning study environment is used to conduct operations planning such as the impact analysis of an outage. Similarly, the engineering studies environment is used to develop historical system performance indices. Study systems are put in place to perform what-if studies. For example, operations planning engineers may examine alternatives other than those produced by the production systems to establish whether they can safely take a transmission line out. The training simulation domain deals with an environment that caters to providing scenarios for training system operators. These scenarios are executed using scripts to provide system conditions that operators face and the actions they need to take. Finally, the quality assurance system is used for testing new applications and upgrades before introducing them to the production system.

Figure 11.3 and Figure 11.4 provide DMS typical displays for the distribution system. These figures indicate detailed distribution feeder topology, loads and other system conditions can be depicted on these diagrams.

11.2.1   System Monitoring

This functionality provides an accurate picture of the system condition in the distribution system. The DMS acquires a significant number (e.g., 100,000) of real-time and near-real-time information about the current status, performance, and loading of distribution system power apparatus. The real-time information would include analogue and status data reporting to DMS SCADA from remote terminal units (RTUs) in substations and feeders once every 4 seconds. The "near real time" information would include equipment measurements and status coming from sources such as smart meters (SMI) that are updated between once every five (5) minutes and once every fifteen (15) minutes. Examples of monitored equipment include

1. Distribution substation transformers and load tap changers
2. Distribution substation switchgear

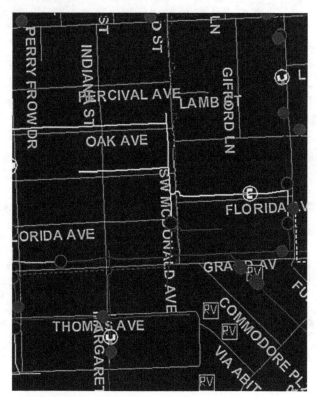

Figure 11.3   A typical DMS feeder display.

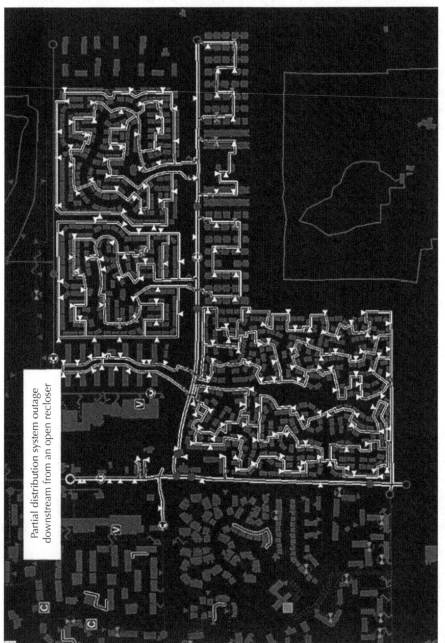

Partial distribution system outage
downstream from an open recloser

Figure 11.4   A typical DMS display including system topology and load.

3. Distribution substation shunt capacitor banks and shunt reactor banks
4. Distribution substation bus meters
5. Field reclosers (overhead, pad-mounted, and installed in underground vaults)
6. Overhead and underground load-break switches with SCADA capabilities
7. Switchable capacitor banks located in the field
8. Field voltage regulators
9. Faulted circuit indicators (FCIs) and other sensors located in the field
10. Distributed generating resources

The transmission of information is conducted using a report-by-exception philosophy, meaning the value will only be transmitted to the DMS when the value changes by a user-definable dead band since the last time the value was transmitted to the DMS.

11.2.2 Decision Support Systems

This functionality provides a rich set of solutions to ensure system reliability, safety, and efficiency. These tools generally use power system information and system models to provide decision support solutions, as well as appropriate control actions. Generally, these applications can be broken into three broad categories: assessment, optimization, and restoration. A more complete list reflecting a typical set of decision support tools in a DMS will be described later in Section 11.3.2.

The applications in each category are shown next, followed by the details of each individual application:

- Assessment applications
  - Distribution system modeler
  - State estimation
  - Load estimation
  - Distribution power flow
  - Short-circuit analysis
  - Short-term load forecasting
- Optimization applications
  - Volt-VAR optimization
  - Optimal network reconfiguration
- Restoration applications
  - Fault location, isolation, and service restoration
  - Tagging

*11.2.2.1 **Distribution System Modeler***   The DMS includes a detailed, up-to-date, three-phase electrical and connectivity model of the electric distribution system as required by the DMS applications such as online power flow and short-circuit analysis. The system model needs to be maintained to ensure that all the network changes are reflected in the model. Also, the model needs to be modified to reflect future permanent additions and changes.

11.2.2.2 State Estimation Similar to state estimation for the EMS, the DMS state estimation tool uses available redundant distribution system measurement values to recreate values for all system state variables. There is a key difference between the transmission and distribution state estimation. The distribution system unlike transmission system is radial. Radial systems comprise a number of feeders originating from a substation supplying electricity to customers without having looped connections. Due to their peculiar properties, these systems need special treatment in their algorithms to ensure solution convergence.

11.2.2.3 Load Estimation Since only major distribution bus loads are measured, a method is needed to distribute the loads to other buses fed by the main station. A load estimation mechanism is required to divide the main bus load among the distribution service transformers. This is done using historical load profiles for each distribution load. The historical load profiles for each bus are deduced from the load mixtures at each bus (e.g., industrial, residential, and commercial) and the load profiles for each load type.

11.2.2.4 Distribution Power Flow The distribution power flow program provides distribution operators with the electrical conditions and power flows in the distribution system to establish abnormal conditions out on the feeders, such as low voltage at the feeder extremities and overloaded line sections. These calculations are initiated in a periodic basis or upon any significant change in the network models or loads. System operators can also initiate a study on demand.

11.2.2.5 Short-Circuit Analysis Short-circuit analysis function calculates the voltages and currents on any of the three phases due to postulated fault conditions with due consideration of prefault loading conditions. The calculated fault currents can be compared against switchgear breaking capabilities or device fault-current limits.

11.2.2.6 Short-Term Load Forecasting The short-term load forecast (STLF) function uses historical load and weather data to forecast the system load automatically every hour, for a period of time such as a week. The load forecasting information supports operations planning applications such as optimal network reconfiguration (ONR). The STLF can be obtained using either of weather forecast method or a similar day forecast method or a combination of both. In the weather forecast method, the load forecast can be obtained using weather forecast properties. A similar day forecast method creates a portfolio of different days in a year having distinct behavior. It then obtains load behavior for these different days in the portfolio using the historical data. The last stage in this method is to categorize each day for which a load forecast is required as one of the distinct days in the portfolio and use the corresponding load behavior.

11.2.2.7 Volt-VAR Optimization Volt-VAR optimization (VVO) determines optimal control actions to minimize an objective function such as load demand or energy consumption while maintaining acceptable voltage and loading at all feeder locations. Examples of control actions include

- Substation transformer control load tap changers (LTCs)
- Three-phase and single-phase voltage regulators located in substations and in the field
- Power generators connected to the distribution system for which supervisory control is available
- Switched capacitor banks located in the field

VVO takes advantage of smart meter measurements to accomplish its functionality. Since the periodicity of the smart meter measurements is large (e.g., 5 minutes), these measurements are called near real time. Figure 11.5 shows the effect of voltage profile before and after VVO control action at the bus where load is being served. VVO generally brings the voltage at the load bus down reducing the demand.

VVO has a feature to abandon its control actions if a component in the sensing devices or control does not function. This feature is called fail safe design.

11.2.2.8 Optimal Network Reconfiguration The ONR function provides the recommended actions necessary to accomplish an objective function without violating any loading or voltage constraints on the feeder. Examples of the objective function could include

- Minimize total electrical losses on the selected group of feeders.
- Minimize the largest peak demand among the selected group of feeders.
- Balance the load between the selected groups of feeders (i.e., transfer load from heavily loaded feeders to lightly loaded feeders).

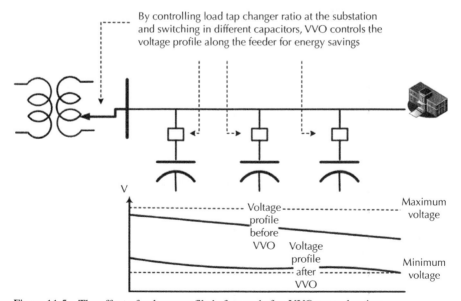

Figure 11.5   The effect of voltage profile before and after VVO control action.

The ONR solution typically provides the recommended switching actions and a switching plan to accomplish these actions for the operator as well as the resulting expected benefits. The specified duration of ONR could span over an operations planning period such as a week. System conditions representing each hour would need to be simulated for optimization.

11.2.2.9 Fault Location, Isolation, and Service Restoration

Fault location, isolation, and service restoration (FLISR) is a restoration functionality when a fault happens. This functionality improves system reliability by reducing the number of customer interruptions and the time to restore the system using controllable devices such as circuit breakers, reclosers, automated line switches, ties switches, fault detectors, and other facilities for monitoring and control.

FLISR consists of the following steps:

- Automatically detect faults
- Automatically determine the approximate location of the fault (i.e., the faulted section of the feeder between two feeder switches)
- Automatically isolate the faulted section of the feeder
- Automatically restore service to as many customers as possible within a very short period of time (e.g., less than one (1) minute following the initial circuit breaker or recloser tripping)

Figure 11.6 shows the status of circuit breakers and fault interruption switches under normal conditions. It shows that both the interruption switches and circuit breaker 1 are closed, whereas circuit breaker 2 is open. In this arrangement, the loads are supplied by substation 1. Let us now assume that there is a fault on the feeder between substations 1 and 2 as shown in Figure 11.7. Figure 11.8 shows the operation of FLISR after the fault on the feeder between substations 1 and 2 indicating the two fault detection and interruption switches open to isolate the fault, and circuit breaker 2 closes to minimize the load loss only to the middle load.

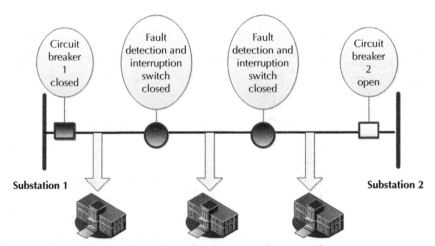

Figure 11.6   There is a fault on a feeder between substations 1 and 2.

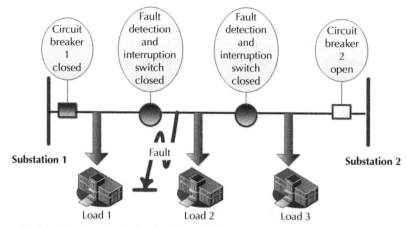

Figure 11.7 A fault on the feeder 2 before FLISR operation.

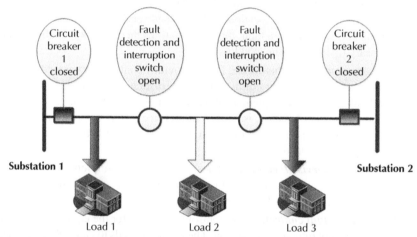

Figure 11.8 The FLISR operation isolates the fault by opening two switches and closing the second circuit breaker 2 to minimize the load loss to load 2.

There are two modes of FLISR operation:

- **Centralized**: In this scheme, the FLISR functionality is executed by the DMS either directly (centralized closed control) or by providing the recommended control action to the operators for their automatic implementation (central supervised control).

- **Decentralized**: In this scheme, the FLISR functionality is accomplished by a peer-to-peer system without DMS intervention. The decentralized actions can be overridden by the operators when needed. Examples include disabling the automated peer-to-peer decentralized operation when a line is being taken out of service to ensure that the automated decentralized operation does not inadvertently energize the line or when the decentralized operation needs to be disabled to change its settings due to changes in network changes.

11.2.2.10   Tagging   DMS has the ability to place tags on any device to inhibit certain remote control commands on the associated facilities in accordance with operating procedures. An example of tags is "Recloser Off Tag," which indicates that automatic reclosing has been turned off for a circuit due to work being performed or due to other reasons.

11.2.3   DMS Control Actions

The DMS is able to control power system apparatus located at distribution substations and field locations. Controlled power apparatus include: circuit breakers; reclosers; motor-operated disconnect switches; substation load tap changers; substation and feeder capacitor banks; substation shunt reactors; voltage regulators; distributed generating facilities; and other high-voltage and medium-voltage equipment.

There are four types of control actions implemented by the DMS:

1. *Direct closed control*: This control action is conducted directly by the DMS with no operator intervention. An example of this kind of control action is the direct closed mode of the fault location, isolation, and service restoration tool. Upon sensing and interrupting a fault, this tool identifies the network reconfigurations necessary to minimize the load loss.

2. *Advisory control*: In this mode, DMS provides control recommendation for implementation. The operator uses his or her discretion on whether the recommendations should be implemented. An example of this kind of control action is placing the fault location, isolation, and service restoration function or VVO function under the advisory control to give the final implementation decision to the operator instead.

3. *Operator supervised controls*: These controls are implemented through EMS by the operators. An example of this control is an operator opening a SCADA-enabled controlled switch.

4. *Operator manual control*: These control actions are for the equipment that the operator cannot remotely operate on. The operator needs to call a plant operator to open a switch because he or she does not have direct control of the switch.

11.3   DISTRIBUTION MANAGEMENT SYSTEM ARCHITECTURE

The high availability requirement feature of the DMS for system operation critical functions forces a double redundant design or a quadruple design similar to the one for EMS shown in Figure 9.2.

11.3.1   Hardware Overview

The DMS hardware overview is similar to the EMS, which generally comprises the following subsystems:

- Data acquisition subsystem
- Computer subsystem

- Man/machine subsystem
- Auxiliary power subsystem/uninterruptible power supply

One of the main differences between the DMS and EMS hardware is the significant number of distribution network components that the DMS needs to monitor and control (i.e., data acquisition subsystem).

11.3.2   Software Overview

The DMS software consists of various decision support tools as well as other administration and support functions needed for DMS to work. Figure 11.9 provides a typical DMS software inventory. A short description for each software is providednext:

1. *SCADA*: Supervisory control and data acquisition system performs data acquisition, alarm processing, man–machine updating, as well as execution of control actions in the field.

2. *Distribution network modeler*: This tool is responsible for maintaining a single-line diagram model of the bulk electricity supply.

3. *State estimation*: This is a mathematical method that uses available power system measurement values to recreate values for all other unknown system state variables.

4. *Remedial action system (RAS)*: These programs are intended to assist the operators in arriving at appropriate remedial control actions to correct for any security violation in the normal system condition and after credible contingencies.

5. *Power flow*: This program provides distribution operators with the electrical conditions and flows in the three-phase distribution system to establish abnormal conditions out on the feeders, such as low voltage at the feeder extremities and overloaded line sections.

6. *Static security (contingency analysis)*: This is a method to identify the system's thermal and voltage violations during normal conditions and after credible contingencies.

7. *Load estimation*: A load estimation mechanism is required to divide the main bus load among the distribution service transformers.

8. *Short-circuit analysis*: This function calculates the voltages and currents on any of the three phases due to postulated fault conditions with due consideration of prefault loading conditions. The calculated fault currents can be compared against switchgear breaking capabilities or device fault-current limits.

9. *Voltage var optimization*: Volt-var optimization (VVO) determines optimal control actions to minimize an objective function such as load demand or energy consumption while maintaining acceptable voltage and loading at all feeder locations.

10. *Fault location, isolation, and service restoration*: Fault location, isolation, and service restoration (FLISR) is a restoration functionality when a fault happens. This functionality improves system reliability by reducing the number of customer interruptions and the time to restore the system using controllable devices such as

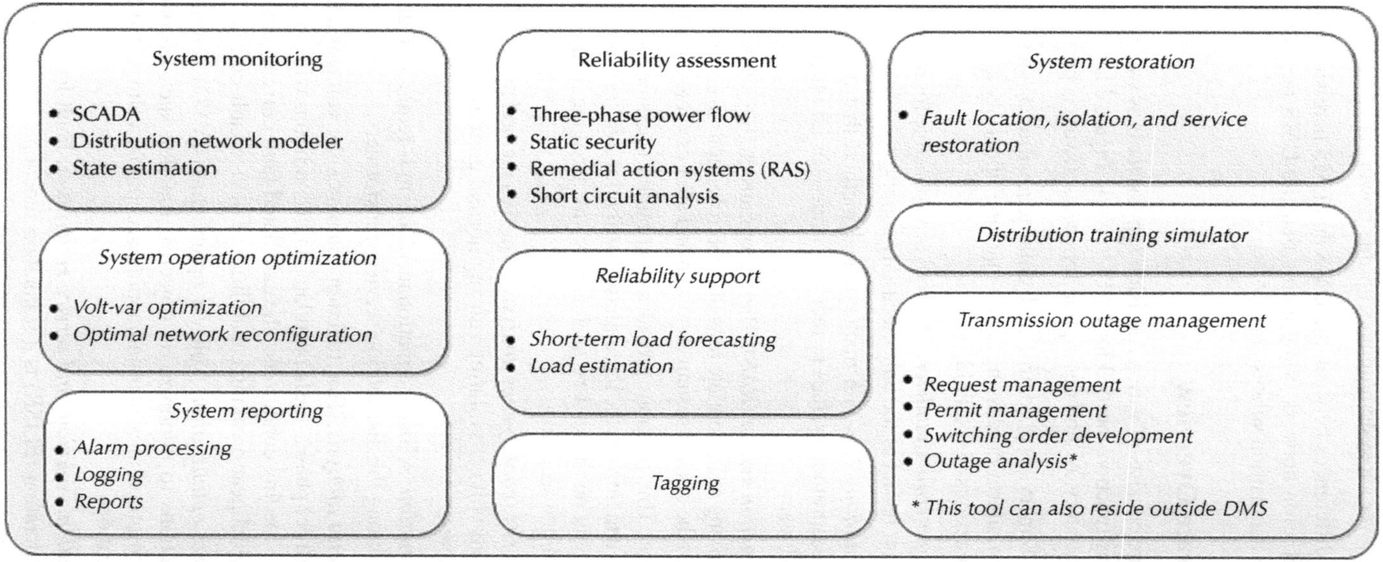

Figure 11.9 Typical DMS software inventory.

circuit breakers, reclosers, automated line switches, ties switches, fault detectors, and other facilities for monitoring and control.

11. *Optimal network reconfiguration*: This function provides the recommended actions necessary to accomplish an objective function without violating any loading or voltage constraints on the feeder.

12. *Short-term load forecasting*: This function uses historical load and weather data to forecast the system load automatically every hour, for a period of time such as a week.

13. *Tagging*: DMS has the ability to place tags on any device to inhibit certain remote control commands on the associated facilities in accordance with operating procedures.

14. *Distribution training simulator*: This tool provides a realistic environment for hands-on dispatcher training under simulated normal, emergency, and restorative operating conditions. The training is based on interactive communication between instructor and trainee with a complete replica of the DMS user interface. The training simulator has its own separate environment.

15. *System reporting*: These tools collect data to produce alarms and the necessary online and historical data summary displays and reports.

11.3.3 Application Integration with DMS

Distribution management system is one of the major software blocks of distribution operation and control. There are other software solutions that the DMS needs to integrate with. To ensure that a solution stands the test of time, the maintenance and sustainment cost for changes during the life span of the solution needs to be minimized. For that reason, current architecture solutions use service-oriented architectures (SOA) to integrate DMS with other system external systems. This solution minimizes the integration work required as a result of changing one of the systems [69].

Moving on the internal architecture of DMS, the roadmap developed by a CIGRE Task Force [69] for the future EMS designs is equally applicable to the DMS. This future design is open and scalable, enabling users to select the best of the breed from different vendors with seamless integration. As explained in Chapter 10, the five salient features constituting the core elements of the proposed design are

- *Service-oriented architectures (SOA)* facilitating third-party integration enabling a change in one component
- *EPRI's common information model compatibility* using consistent information models
- *Built-in security* to avoid piecemeal patching
- *Platform independence* ensuring that solutions can work on any platform
- *Unified graphical interface* decoupling the tool from the display design

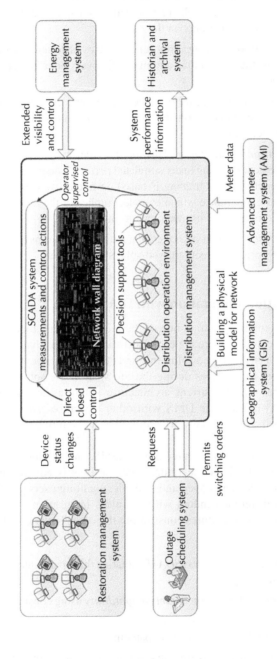

Figure 11.10 DMS interactions with other systems.

Figure 11.10 shows the systems interacting with DMS and their information exchange:

1. Restoration or outage management system (OMS)
2. Outage scheduling system (OSM)
3. Geographical information systems (GIS)
4. Advanced meter management system (AMI)
5. Historian and data archival system
6. Energy management system (EMS)

A description of the functionality of the external systems interacting with the DMS appears in the following subsections.

11.3.3.1 *Restoration or Outage Management System (OMS)* This tool provides an environment for information exchange with customers collecting the customer outage conditions and assisting with the system restoration. The major functions in an OMS include [78]

- Locating the system terminal points (e.g., location of fuse or breaker) separating energized system from the nonenergized system resulting from the failure.
- Organizing and ranking restoration tasks using criteria such as the criticality of the emergency, size, and duration.
- Estimating and planning for the required crews and managing the crews.
- Providing information on the outages and impacted customers, estimating the restoration times.

A modern outage management system is a detailed network model of the distribution system. It uses the geographical information system (GIS) as the source of this network model to establish the physical location of the fault. It also integrates with the DMS SCADA to keep its system electrical model up-to-date so that it can accurately make outage predictions and keep track of which customers are out or restored. By using this model and by tracking which switches, breakers, and fuses are open or closed, network tracing functions can be used to identify every customer who is out, when they were first out, and when they were restored. Tracking this information is the key to accurately reporting outage statistics.

11.3.3.2 *Outage Scheduling System (OSM)* This tool provides an enterprise solution that provides the following functionalities:

- Submission of outage request by field staff.
- Assessment of outage impact by operators.
- Development of switch orders by operators. Switching orders consist of a list of actions that are needed to perform switching such as opening/closing various types of switches, blocking, grounding, and tagging.
- Execution of the switching orders by operators.
- Development of the permits by operators for field staff.
- Collection and reporting of outage information.

This tool will be described in more detail in Chapter 12 discussion of outage scheduling systems for both distribution and transmission systems.

11.3.3.3 Geographical Information System (GIS) Distribution network modeler includes the dynamic network model, which reflects the status of switches and circuit breakers. This model is supplemented with the GIS model of the network for distribution operators to get a physical sense of the network model. The GIS supplies up-to-date network connectivity information as well as electrical and physical characteristics information of distribution system components. The GIS supplies physical details about the distribution lines, including line length, construction type, conductor arrangement and spacing, wire size, number of conductors, circuit type (overhead line, underground cable, etc.), and other such information. This tool is responsible for creating a single-line diagram model of the bulk electricity supply. The interface between the GIS and the DMS provides a means for data exchange from the GIS database to the DMS to reflect the initial physical aspects of the network as well as changes that takes place in time.

11.3.3.4 Advanced Meter Management System (AMI) This interface allows the transfer of near real-time (e.g., every 5 minutes) meter information to the DMS for additional visibility.

11.3.3.5 Energy Management System (EMS) This interface allows the exchange of monitoring and control between DMS and EMS systems. For example, monitoring of some transmission equipment can be transferred to DMS for better visibility. The control of some controls in the distribution system such as opening a switch can also be transferred to the EMS and vice versa.

11.3.3.6 Historian and Data Archival System The historian and data archival system provides a means for storing system conditions and their retrieval for system analysis.

QUESTIONS AND PROBLEMS

11.1. What are the business justifications for a DMS?

11.2. Compare the number of equipment devices in DMS and EMS.

11.3. What are the main categories of DMS functionality?

11.4. What are the three categories of DMS applications?

11.5. What are the domains that the DMS functionality is offered in?

11.6. Describe the functionality of VVO.

11.7. Describe the functionality of localized and centralized FLISR.

11.8. Describe the functionality optimal network reconfiguration.

11.9. Describe the two categories of DMS measurements.

11.10. Describe the systems that DMS interfaces with.

CHAPTER 12

EVOLVING POWER SYSTEM OPERATION SOLUTIONS

12.1 INTRODUCTION

This chapter covers the evolving and the state of the art in power system operation solutions. These solutions, which cover different domains of transmission and distribution operation, have different levels of maturity and will be available for production deployment in different time frames. To provide context on the evolution maturity, the solutions presented in this chapter are categorized under the following three availability time frames:

1. Readily available
2. Mid term (e.g., 5 years)
3. Long term (e.g.,10 years)

12.2 EVOLVING OPERATION SOLUTIONS

12.2.1 Online Transient Stability

As described in Chapter 5, offline transient stability limits are derived by conducting a large number of offline time-domain simulations reflecting system dynamic response following a number of critical contingencies for system normal (i.e., all the elements in service) and the system with one element out of service. Studying conditions with one element out of service is necessary to provide a margin for the uncertainty of operating conditions. Because of maintenance and forced outages, the actual system is rarely in a state with all elements in service. The problems associated with offline methods are

- Since the prevailing operating conditions are not known ahead of time, the worst operating condition for the period of study (e.g., winter season) is selected. This creates a degree of conservatism in deriving the limit.
- Since the prevailing network conditions are not known ahead of time, the system normal model and the system model with one critical element out of service is considered in the simulations. This assumption may prove to be very wrong

Practical Power System Operation, First Edition. Ebrahim Vaahedi.
© 2014 The Institute of Electrical and Electronics Engineers, Inc. Published 2014 by John Wiley & Sons, Inc.

depending on the prevailing system network model. If the system has suffered a major contingency, the assumption creates a degree of nonconservatism.

- Since the prevailing system network model is not known ahead of time, the contingencies considered could be invalid, creating a degree of risk.
- Offline limit derivation is a very engineering-intensive task.

As an example, let us assume that we want to derive the winter season transient stability limits for a 1000 bus system considering 50 critical contingencies. Let us also assume that we would like to consider 10 major system outages that can happen due to system maintenance requirements. The offline calculations use the following steps to calculate the limits:

- Use the worst system condition scenario for studies (e.g., winter peak).
- Derive the transient stability limit using the "system normal" scenario (no element out of service) for the 50 critical contingencies.
- Derive the transient stability limit using the "system with one outage element" (e.g., 10 scenarios) for the 50 critical contingencies.
- Archive the results and use them to set the flow gates to ensure system stability.

This process indicates that we need to conduct 550 transient stability limit calculations (11 base-case scenarios each considering 50 critical contingencies), which are all based on the worst-scenario condition (e.g., winter peak).

Online transient stability, on the other hand, uses a real-time snapshot of the system prevailing condition every few minutes (e.g., 5 minutes) and calculates the limits for 50 critical contingencies. So while this process gets repeated every few minutes (e.g., 5 minutes), the number of limit calculations is only 50. For these calculations, there is no need to conduct limit calculations for other network model conditions (one element out of service) because the prevailing network condition is being considered. Furthermore, the results are more accurate because the prevailing loading conditions are used as opposed to the worst scenario condition (e.g., winter peak).

The aforementioned discussion indicates that offline limit calculation creates approximate limits that can either be too conservative or nonconservative. If the limits are conservative, the existing transmission line capacity is not being used fully, creating inefficiency. On the other hand, if they are nonconservative, the limits derived pose the risk of system instability. Furthermore, the process of offline derivation is a very inefficient process using a lot of engineering time.

For these reasons, the time domain-based online transient stability method [79–81] has been developed. Online transient stability uses the prevailing system and network conditions to establish transient stability limits and remedial actions needed for different critical contingencies on a cyclic basis (e.g., every 5 minutes). These methods use time domain simulation engines similar to those used for offline simulations; their biggest strength is their ability to simulate all the models and process consistent with offline limit derivation. Figure 12.1 shows a typical operating region nomogram provided by an online TSA tool. Operators ensure that the operating point does not violate the operating region.

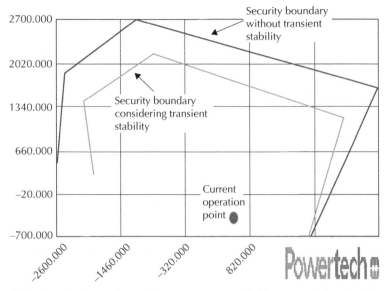

Figure 12.1   A typical operating region nomogram provided by an online TSA tool.

Since it is impossible to consider all the probable contingencies in an online transient stability calculation, a fast method is needed to screen and rank the contingencies. These methods normally fall under the following categories:

1. Energy methods [82–84]
2. Single-machine infinite bus methods [84]
3. Pattern recognition methods [85]

The energy methods are based on the concepts of defining an energy function for the system that reflects the dynamics of the system. Using the defined energy function, the stability boundary of the system can be determined directly. The strength of these methods is their speed. Their drawback is the compromised accuracy making them ideal for contingency screening and ranking [82,84].

Single-machine infinite representation [1] entails reducing the system to one machine from which the limits can be derived using equal area criteria. Again, the strength of this method is speed, and their drawback is its accuracy. Again, this method is suited for contingency screening and ranking [84].

The pattern recognition methods use a large number of samples and try to obtain a common pattern using some key parameters, which would then be used to classify different new samples. These methods, similar to other approximate techniques, enjoy low computation time but lack accuracy [85].

An online transient stability method requires advanced applications for contingency screening and ranking as well as limit calculations. It also requires a flexible rule engines to easily customize business rule. A complete online transient stability solution is expected to be available within the mid-term time frame.

12.2.2 Online Voltage Stability

As described in Chapter 6, offline voltage stability limits are derived by conducting a large number of offline time-domain simulations reflecting dynamic system performance following a number of critical contingencies for the "system normal" (i.e., all elements in service) and the "system with one element out of service."

Similar to offline transient stability assessment, offline voltage stability assessment suffers from the following features:

- Since the prevailing operating conditions are not known ahead of time, the worst operating condition for the period of study (e.g., winter season) is selected. This creates a degree of conservatism in deriving the limit.

- Since the prevailing network conditions are not known ahead of time, the system normal model and the system model with one critical element out of service is considered in the simulations. This assumption may prove to be very wrong depending on the prevailing system network model. If the system has suffered a major contingency, the assumption creates a degree of nonconservatism.

- Since the prevailing system network model is not known ahead of time, the contingencies considered could be invalid, creating a degree of risk.

- Offline limit derivation is a very engineering-intensive task.

Online voltage stability tools access the prevailing system state and topology conditions available in real time from the energy management system. This allows the tool to only study the appropriate system condition avoiding unnecessary scenarios and conservative assumptions. Voltage stability assessments are then only conducted reflecting the dynamic performance of the system after the critical contingencies for the prevailing system condition. Another feature of online voltage stability is that their calculated limits use less uncertainty and need less margin of safety.

In recent years, online voltage stability tools have been developed that calculate the voltage stability margin on a cyclic basis (e.g., 5 minutes) and recommends remedial actions necessary to meet the criteria [86,34]. It is important to note that online voltage stability methods calculate voltage stability margin in a consistent manner to offline voltage stability with the same process and modeling accuracy.

Figure 12.2 shows an operating region provided on EMS by BC Hydro's online voltage stability program. Operators ensure that the operating point does not violate the operating region.

Since it is impossible to consider all the probable contingencies in an online voltage stability calculation, a fast method is needed to screen and rank the contingencies [34,39]. The tool should also be able to recommend preventive or corrective control actions to improve voltage security of the system if the system is found to be voltage insecure after any credible contingency. These requirements extend the functions of the online voltage stability tool to

- Contingency selection and screening
- Voltage security evaluation
- Voltage security enhancement

An online voltage stability solution is readily available for deployment.

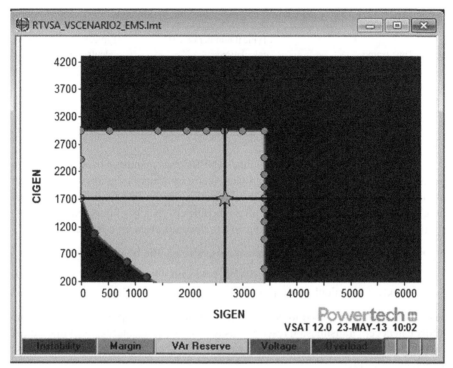

Figure 12.2   BC Hydro's online voltage display.

12.2.3 Total Transfer Capability Calculator

Total transfer capability (TTC) is defined as the amount of electric power that can be transferred over a path or interconnected transmission network in a reliable manner. In other words, the TTC of an intertie is defined as the maximum power flow fulfilling the following conditions:

- Precontingency conditions within voltage profile and thermal limits.
- Postcontingency conditions within postcontingency voltage profile and postcontingency thermal limits.
- Precontingency and postcontingency conditions within respective precontingency and postcontingency voltage stability limits ensuring that the prevailing system condition will not suffer from voltage instability under normal conditions or under conditions resulting from any $(n-1)$ critical contingency.
- Postcontingency conditions within transient stability limit ensuring that the system will not suffer from transient instability following any $(n-1)$ critical contingency.

The TTC of various paths on the transmission system must be calculated in order to operate the system reliably and to determine the available transfer capability (ATC). The ATC is the amount of transmission that can be made available to provide transmission service.

TABLE 12.1 TTC calculation time frames and calculation cycle

Time frame	Typical number of TTC calculations	Typical calculation cycle
Real time, hourly	12	Hourly
Prescheduling, hourly	168	Daily
Prescheduling, daily	30 on-peak and off-peak	Weekly
Prescheduling, weekly	52 on-peak and off-peak	Weekly

The North American Electric Council (NERC) has mandated [87] the utilities to publish ATC on their interties to facilitate electricity transactions between different utilities. Utilities need to publish this information on the two time frames of "real time" and "prescheduling." The real-time schedule covers a few hours (e.g., 12 hours) starting at the next hour. Prescheduling covers three different time frames of

- Hourly TTC calculations for about a week (e.g., 168 hours)
- Daily TTC calculations (peak and off-peak) for about 1 month (e.g., 30 days)
- Weekly TTC calculations (peak and off-peak) for about 1 year (e.g., 52 weeks)

These calculations are used to post the ATC on interties for up to a year so that the utility can accept transmission reservations for the use of its interties on a first-come, first-served basis. These calculations are repeated in scheduled periods such as those shown in Table 12.1.

To calculate the TTC values properly, the following steps must be followed:

1. Develop the system condition base-case (about 330) reflective of the scenarios to be studied. This would need load forecast, generation forecast, generation dispatch and outages.

2. Simulate system conditions with $n - 1$ critical contingencies determining

   a. Static security limits, ensuring that the thermal limits and voltage profiles are respected

   b. Transient stability limits

   c. Voltage stability limits

3. Calculate the TTC limits using the smallest of the three limits.

Figure 12.3 shows BC Hydro's ATC calculation display and typical results for the real-time time frame (12 hours).

NERC recognizes that there are uncertainties associated with the TTC calculations arising from a number of factors such as load forecast, transmission topology, and generation dispatch, and that the effect of the uncertainties needs to be estimated and taken into account [88,89].

Once the TTC values are calculated, they are posted on the transmission market site called OASIS (Open Access Same Information System). In compliance with FERC open transmission access rules, the transmission customers can make reservations on the transmission capacity up to a year on a first-come, first-served basis.

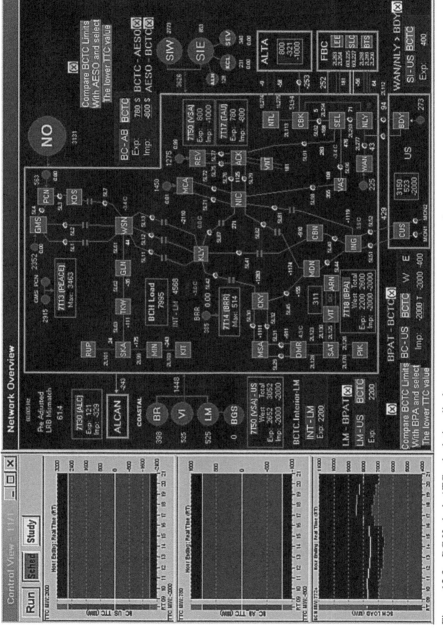

Figure 12.3   BC Hydro's ATC calculation display.

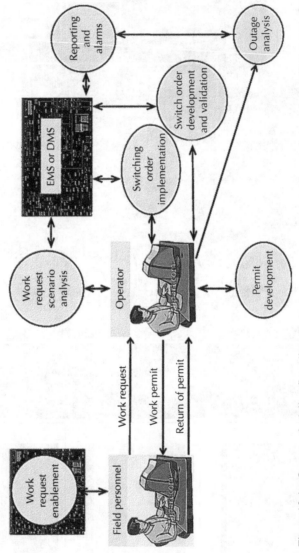

Figure 12.4 Outage scheduling flow diagram.

An accurate TTC method requires online transient stability and online voltage stability engines. This solution is expected to be available within a mid-term time frame.

12.2.4 Transmission Outage Scheduling System

Transmission outage scheduling systems provide a collaborative tool to perform the tasks associated with planned outages. It also provides a medium to document planned and unplanned outage events for future reliability analysis. As shown in Figure 12.4, a transmission outage scheduling system provides the following functions:

1. **Enabling requests**: This function enables the field personnel to send in outage requests, validate outage request information, and deny the outage if there is information insufficiency. This is normally done using a tabular screen or a one-line diagram.

2. **Planning/reporting**: This task enables the control room personnel to assess the reliability impact of the outage and plan the outage when it is feasibly possible. This ideally uses an advanced reliability assessment solution environment such as those existing in an EMS or EMS solutions.

3. **Developing and implementing switching order**: In this task, the control room staffs develop the switching order to isolate the equipment. This can be done automatically using the dynamic network connectivity model that will be automatically created using the isolation points of an outage. The switching order is then implemented automatically from the control room or by contacting the field staff if automatic switching is not possible.

4. **Implementing permit and returning permit**: Following the execution of the switching order, the operators develop a protection permit. As part of this process, first the operator places tags on isolation points of the outage on the mimic diagram to ensure that the control room staffs do not inadvertently place the outage back in service. Then the operator sends the permit to the field workers so that they can start working on the outage. Once the work is completed, the field workers return the permit to the operator so that he or she can start restoring the system.

5. **Developing and deploying return switching order**: Once the field workers have sent back the permit, the system operator needs to develop a return switching order so that the system can be put together. The tool would be used to create the switching order that will either be used by the operator for automatic switching or relayed to the field personnel for manual switching. If the switching order is executed by the field personnel, the status will be reported to the operator.

6. **Events and outage analysis**: Normally there are a number of unscheduled outages and events that take place in the system. The operator needs the functionality to document these events for restoration and reliability analysis.

A transmission outage scheduling system solution is expected to be available within the mid-term time frame.

12.2.5   Synchrophasor Systems

Synchrophasor systems provide accurate, high-speed grid measurements such as voltage, current, and real power at a very high speed. A typical phasor measurement unit (PMU) sampling rate of 30 per second can be compared with that of a RTU of one measurement every 2 seconds. The increased PMU measurement sampling rate provides tremendous visibility compared to the conventional RTU measurements. This situation has been anecdotally compared to the visibility afforded by a MRI versus that of an X-ray machine [90].

Each PMU measurement is time-stamped according to a common time reference. Time stamping allows the measurements from different locations to be synchronized and combined, providing a precise and comprehensive view of the entire interconnection.

These measurements open a new domain for visibility and control in power systems. The high-speed visibility and control has the potential to provide system operators with improved situational awareness of grid problems. System operators can identify, visualize, and analyze system vulnerabilities and disturbances as they are developing in real time. With this capability, operators can take timely remedial actions to stabilize the power system and avoid widespread system disturbances. So far, synchrophasor systems have provided practical applications [90–92] in the following four broad areas:

- **Wide-area monitoring**: Phasor data are collected and fed into processing applications that provide better situational awareness to grid operators, enabling them to develop remedial actions to protect system reliability.

- **Real-time operation**: Phasor data are used in different real-time applications to enhance system reliability. Examples include interarea oscillation monitoring, enhanced state estimation, voltage stability monitor, and improved remedial action schemes.

- **Power system planning**: Phasor data can be used for calibrating system models used in simulations to enhance the accuracy of the system performance assessment.

- **Disturbance analysis**: Phasor data are used to analyze the system performance following major disturbances. The archived PMU data representing the conditions across the grid can be analyzed to determine the sequence of events and the causes of disturbances.

Figure 12.5 shows the deployment of PMU as of November 2012. As a result of US government stimulus funding, PMU adoption has increased in recent years. This investment will result in the deployment of over 1000 PMUs by 2014 networked to provide coverage of wide-area transmission system. Also, North American Synchrophasor Initiative (NASPI) [92], which has been established to improve power system reliability and visibility through wide-area measurement and control, continues to promote the technology maturity and deployment moving forward. Despite all the heightened degree of research and development in recent years, PMU technology has ways to go to capture its promised potential. Gartner, the largest technology benchmarking company in North America, has placed this technology in their ranking category of "Trough of Disillusion" [93]. This category belongs to the technologies that they have not yet fulfilled their expected potential.

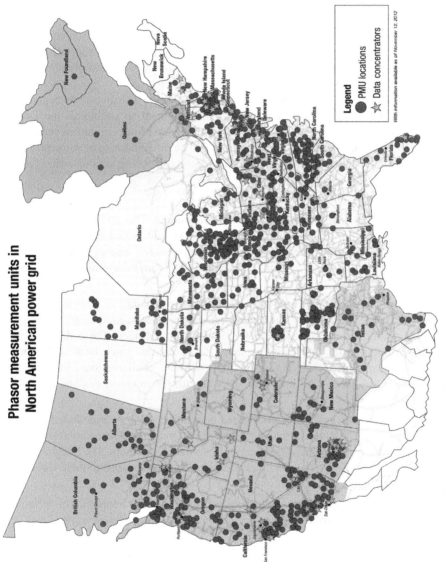

Figure 12.5 PMU installations in North America as of November 12, 2012.

While basic synchrophasor applications are readily available for deployment, it is expected that more mature synchrophasor applications will be available in the mid-term time frame.

12.2.6 Distribution Automation

Distribution automation (DA) enables the utility to monitor distribution equipment remotely, gather system and equipment information, and take appropriate control actions with or without operator intervention to respond to system abnormalities, adjust to dynamic loads, or to meet optimization objectives.

DA can be also thought of as the compliment to the distribution management system by providing the communication and equipment to enable all the DMS advanced functionality. For example, DA uses intelligent power equipment and intelligent electronic devices (IEDs) in substations and on distribution to monitor, protect, and control the distribution network such as locating, isolating, and restoring faults automatically.

DA includes functions such as

1. *Optimizing the distribution system performance*: This includes deployment of distribution management system and all its advanced system reliability and safety applications such as fault management, energy efficiency and loss reduction, and capacity utilization by load balancing and demand management.

2. *Advancing protection, control, and monitoring capabilities*: This includes the use of advanced IEDs and intelligent power system equipment that allow teleprotection setting and remote configuration changes. These devices and equipment ensure safety while enabling the distribution grid to be more dynamic and adaptive to changing distribution network configurations, intermittent energy resources, and load diversities due to, for instance, demand response (DR) and PEV.

3. *Improving integration of distributed generation (DG) or distributed energy resources (DER)*: Distributed generation [94] generates electricity from many small energy sources located very close to the load centers. The integration of these emerging energy resources necessitates improvements to their visibility, forecasting, dispatch, and control.

4. *Improving asset investment planning, maintenance and operation*: This includes updating the distribution planning, maintenance, and operation guidelines to accommodate the DA.

5. *Developing the DA communication infrastructure*: This includes the development of a communication solution to enable DA to deliver the service level in terms of latency, availability, and security.

The DA investment is huge since it involves the installation of devices all over the distribution system. It is expected that mature DA applications will be available in the mid-term time frame.

12.2.7 Dynamic Thermal Rating Systems

By design, transmission lines have a maximum thermal rating. This limit represents the maximum electrical power or current that can be transmitted through the line safely. Violation of the thermal limits may cause the conductors to sag too closely to the ground or damage the conductors by overheating.

To obtain the maximum current that can flow through a line, a heat balance equation such as the one given next should be used [95]:

$$Q_{\text{loss}} + Q_{\text{sun}} = Q_{\text{rad}} + Q_{\text{con}} + mC_{\text{p}} \frac{dT}{dt}$$

where, Q_{loss} is the equipment losses heat, it is a function of current I; Q_{sun} is the heat input from the sun that can be measured or calculated; Q_{rad} is the radiated heat away from the line, it is a function of the temperature; Q_{con} is the convection heat moving away from the line, it is a function of direction and speed of wind; $mC_{\text{p}} dT/dt$ is a heat storage term with a steady state value of zero where m is the mass; C_{p} is a constant; and T is the temperature.

This equation can be solved to obtain a dynamic thermal rating, which is the maximum allowed current, by measuring the following values:

1. Wind speed measured by an anemometer
2. Wind direction
3. Air temperature
4. Solar heat intensity
5. Conductor parameters

A variation to this method is to measure the actual conductor temperature in addition to the aforementioned measurements.

This technology can significantly help to increase the transmission thermal limit. The use of this technology, however, has been limited due to two different factors:

• Despite all the development efforts so far, there are no general models that can be applied to different types of overhead and underground cables. Models need to be developed for different transmission lines based on their type and their ratings.

• Instrumentation requirement is not trivial considering these measurements need to be repeated along the transmission line.

Due to the required technology enhancements to move this solution from an R&D stage to a production environment, it is expected that dynamic thermal rating solutions will be available in the mid-term or long-term time frame.

12.2.8 Distributed Energy Resources

Historically, electricity has been generated in large remote centralized facilities such as fossil fuel, nuclear, and hydropower plants to take advantage of their favorable economies of scale. These plants in turn transmit through high-voltage transmission and distribution systems. These systems, while enjoying favorable generation cost, need to contend with the following unfavorable factors:

- Transmission and distribution cost
- Transmission and distribution losses
- Environmental implications resulting from generation, transmission, and distribution

DER or DG [94] generate electricity from many small energy sources located very close to the load centers. Since it allows collection of energy from many sources, it results in lower environmental impacts and improves security of supply. While the generation cost of DG is more expensive than conventional sources on a kWh basis, the additional cost is expected to decline as technology improves and more value is placed on the positive attributes of DER, such as lower negative externalities (e.g., environmental impacts).

DER system uses a variety of renewable and nonrenewable technologies such as combined heat power (CHP), fuel cells, microturbines, wind energy, biopower, reciprocating engines, photovoltaic systems, and storage systems. Nonrenewable DER technologies such as reciprocating engines, combustion and microturbines, and fuel cells use natural gas or some other types of fuels.

One of the major challenges associated with DER is the management of the distributed resources to ensure that they are all coordinated to optimally balance generation and load. Similar to the conventional economic dispatch and unit commitment solutions, a DER decision support system is needed to coordinate the operation of the distributed system. For example, the DER decision support system can determine the operation of the storage system to eliminate system peaks or enable electric vehicles to be charged in an optimized way.

The technology needed for the integration and management of DER is expected to be available in the mid-term time frame.

12.2.9 Demand Response

The definition of DR adopted by the Federal Energy Regulatory Commission (FERC) [96] is stated as follows:

Changes in electric usage by end-use customers from their normal consumption patterns in response to changes in the price of electricity over time, or to incentive payments designed to induce lower electricity use at time of high wholesale market prices or when system reliability is jeopardized.

DR enables customers to contribute to energy load reduction during times of peak demand or high market prices. The motivation for DR stems from the fact that electrical generation and transmission systems are generally sized to correspond to peak demand (plus margin for forecasting error and unforeseen events); therefore, lowering peak demand reduces overall plant and investment cost requirements. DR generally results in the shift of the peak demand time to the low demand periods.

DR can be implemented for all types of customers such as large and small commercial as well as residential customers. Often times, it is implemented through the use of dedicated control systems to shed loads in response to a request by a utility or market price conditions.

In general, there are three DE categories:

- *Emergency DR*: The use of this DR allows service interruptions during times of supply scarcity. This type of demand response allows the utility to shed targeted loads as opposed to indiscriminate load shedding.

- *Economic DR* is employed to allow electricity customers to curtail their consumption when the productive or convenience of consuming that electricity is worth less to them than paying for the electricity.

- *Ancillary services DR* consists of a number of specialty services that are needed to ensure the secure operation of the transmission grid that have been traditionally provided by generators [97].

It is expected that mature DR solutions will be available in the mid-term time frame.

12.2.10 Microgrid

A microgrid is a localized small-scale system at the distribution level containing its own generation and load and is designed to be self-sufficient by using advanced monitoring and control solution [94]. Microgrids can connect to a traditional centralized grid for back-up supply or can operate in a remote community, which is nonintegrated to the centralized grid. The microgrid generation includes DER such as storage systems, wind, and solar that are geographically dispersed to increase the security of supply.

There are a number of challenges associated with the microgrid operation:

1. The first challenge is to develop a monitoring and control systems that uses the flexibility afforded by generation and load in a microgrid to make it operate in a self-sufficient manner under normal condition and under $n - 1$ contingency condition.

2. The microgrid's self-sufficiency control operation should be harmonized with the main system's AGC control. If the microgrid is connected to the main system grid, the microgrid's self-sufficiency control should work with the AGC to balance its generation and load. If the microgrid is islanded or does not have a connection with the grid, its own self-sufficiency control should act to maintain the microgrid's frequency by balancing generation and load within the microgrid.

3. Grid operators face two challenges when islanding or restoring microgrids. To ensure reliability, the grid operator should have a good forecast of the generation load imbalance in the microgrid. Another challenge that is more important is the safety of the field people during islanding and during restoration of the microgrid. Procedures need to be developed to ensure that the crews' safety is not compromised.

It is expected that mature solutions for microgrids will be available in the long-term time frame.

12.2.11   Real-Time Posturing and Control

In operating a power system, there are occasions when the operator finds an unacceptable prevailing operating condition. Unacceptable conditions can results from violations either under the normal conditions or following $(n-1)$ contingencies. The violation could be as a result of either of the following limits:

1. Thermal limit (under normal condition or $n-1$ condition)
2. Voltage profile limit (under normal condition or $n-1$ condition)
3. Voltage stability limit (under normal condition or $n-1$ conditions)
4. Transient stability limit ($n-1$ conditions)

Operators need to rectify the violation using the most effective control for the phenomenon at their disposal. Presently, while there are tools that handle some of the violations [67], there are no tools that handle all the violations listed earlier. Operators normally resort to their own experience to fill the vacuum for a comprehensive tool.

It is expected that real-time posturing and control solutions will be available in the long-term time frame.

12.2.12   Critical System Application and Facilities Heartbeat

The Outage Task Force Final Blackout Report concluded [98] that one of the initiating causes of the 2003 blackout was the loss of the critical monitoring tool by the utilities involved. In effect, the utilities had lost the situation awareness of the degraded system conditions. This situation was even further compounded due to the fact that the operators did not know the monitoring tool displays were indeed frozen providing a false sense of security. As a result of this finding, NERC's Real-Time Tools Best Practices Task Force in its 2008 report [98] called for a critical applications and facilities monitoring tool that will reside on an independent system to track the status and availability of real-time tools. This tool ensures that the information provided by the critical applications and their infrastructure tools is current and continuously available to operators and technical support staff. It is important that the monitoring tool resides on an independent system so that it is not impacted by the critical application and their infrastructure.

The idea for a critical application monitor can be extended to other areas of operation such as market applications and distribution management system. So, it is important to think of this application as a central heartbeat that can monitor independently the critical processes, their applications, and infrastructures.

It is expected that critical system application and facilities heartbeat solutions will be available in the long-term time frame.

12.2.13   Probabilistic Limit Calculations

Currently, utilities use deterministic criteria for establishing their transmission operating limits. In other words, the stochastic nature or the variability of many system parameters are not considered. NERC requires that utilities obtain their operating limits using the $n-1$ criteria in ensuring that the system operates within acceptable operating limits after any single contingency.

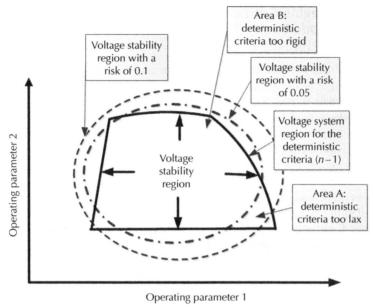

Figure 12.6   Deterministic and probabilistic calculations of voltage stability regions.

It would be useful for the operators to know the amount of risk they may be taking by observing or violating the deterministic limits. Alternatively, the operators may decide to use a more restrictive limit if the deterministic limit still exposes the utility to unacceptable risks.

Figure 12.6 shows a typical voltage stability contour providing the operator with the acceptable region of operation using the $n-1$ criteria. Other contours derived show the operating region with a specified risk of 5% or 10%. A risk of 5% would mean that there is a 5% likelihood that the system could suffer from voltage instability if the operating condition remains within 5% risk region due to other lower probability events that have to not been considered. Now let us assume that the system operating condition is in area A of the voltage stability region. In this area, the deterministic criteria are too lax because the voltage stability region with a risk of 0.1 is more limiting than the deterministic criteria. So the operator should be cautious not to take any further risk. On the other hand, in area B of the voltage stability region, the deterministic limit is too rigid because it is more limiting than the one contour with a risk of 5%. So in this area the operator may decide to take more risk even if it is violating the deterministic criteria.

Probabilistic methods unlike deterministic methods consider the random nature of many system parameters and operating conditions and can come up with the stability limits and their system risk associated with observing the derived limits. Methods to derive the probabilistic limits [99] have not yet matured sufficiently and more research and development is needed to develop robust solutions.

It is expected that probabilistic limit calculation solutions will be available in the long-term time frame.

12.2.14 Managing Critical Operations Knowledge: Operations Code Book

Further Reading

In electric utilities worldwide, control room operators rely on their extensive experience, knowledge of the system, and its current conditions as well as the decision support information from a number of real-time systems (such as EMS, DMS, and OMS) to make critical decisions. These decisions are often governed by a number of rules, procedures, constraints, and business practices specific to the utility and the operating environments. To appropriately balance the required functions with timely response, operators often need to develop business processes and detailed step by step procedures that articulate the actions they need to take under different system conditions. In a number of utilities, these procedures and processes are stored in a book called the "operations code" book, which is made available to operators in a hardcopy form or in an electronic form. While the code book has proven to be a necessary enabler for operators in performing their jobs, the process of finding procedures for executions remains slow and at times ineffective. During emergency operating conditions such as complete or partial blackouts, quick access to accurate and complete information is critical. Hence, making the information in the code book available in an efficient and timely manner is valuable. Technology-based solutions to expedite the process of consulting the operations code book can significantly enhance system reliability, safety, and operations excellence. Solutions are needed optimizing the access and consultation of the code book based on the operating situation and the operator's specific need.

It is expected that managing critical operations knowledge solutions will be available in the mid-term time frame.

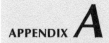

PRELIMINARY CONCEPTS

A.1 INTRODUCTION

This section reviews some basic concepts that will be used throughout the book. It covers phasor representation, active and reactive power derivation, per-unit systems, network modeling, and development of matrices representing the network.

A.2 PHASOR REPRESENTATION

Phasor representation is an approach that simplifies the calculation by converting a sinusoidal function to a vector. Consider a general sinusoidal function $f(t)$:

$$f(t) = F_{max} \cos(\omega t + \phi) = \sqrt{2} F \cos(\omega t + \phi) \tag{A.1}$$

There are three parameters associated with this function, F_{max}, the maximum value, ω, which is the frequency, and ϕ, which is the phase angle. F is called the RMS value (root mean square).

This function can be written in a different form using Euler's equation:

$$e^{j\Theta} = \cos\Theta + j\sin\Theta \tag{A.2}$$

Using this equation, it can be shown that $f(t)$ can be written as

$$f(t) = \sqrt{2}\,\mathrm{Re}\left[Fe^{j(\omega t + \phi)} \right] = \sqrt{2}\,\mathrm{Re}\left[Fe^{j\omega t} e^{j\phi} \right] = \sqrt{2}\,\mathrm{Re}\left[F_p e^{j\omega t} \right] \tag{A.3}$$

where $F_p = Fe^{j\phi}$ is called the phasor representation of Equation A1.3 as shown in Figure A.1.

It can be shown that two sinusoidal functions with the same frequency can be added up by vector additions of their phasors. Kirchhoff's voltage and current laws can be applied using phasors as follows:

Practical Power System Operation, First Edition. Ebrahim Vaahedi.
© 2014 The Institute of Electrical and Electronics Engineers, Inc. Published 2014 by John Wiley & Sons, Inc.

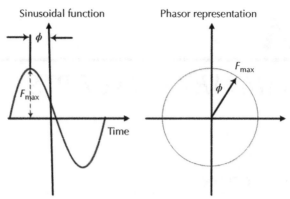

Figure A.1   Phasor representation of a sinusoidal function.

Kirchhoff's voltage law: The phasor addition of voltage drops in a closed circuit is equal to zero.

Kirchhoff's current law: The total phasor addition of all currents coming to a node is equal to zero.

In power system studies, phasors are used for representations of currents, voltages, real power, and reactive power. Using the phasor representations, the real and reactive power in a single circuit can be obtained by

$$P = |V| \cdot |I| \cdot \cos\phi \qquad (A.4)$$

$$Q = |V| \cdot |I| \cdot \sin\phi \qquad (A.5)$$

where ϕ is the phase angle between the current and the voltage. Equations A1.4 and A1.5 can be put in a complex form:

$$S = P + JQ = |V| \cdot |I| \cdot \cos\phi + j|V| \cdot |I| \cdot \sin\phi = V \cdot I^* \qquad (A.6)$$

where I^* is defined by

$$I^* = |I| \cdot e^{-j\Theta i} \qquad (A.7)$$

where Θi is the phase angle of the current.

We can also state S, the complex power, in terms of Y or Z (admittance or impedance) as

$$c_{ij} = a_{ij} + b_{ij} \qquad (A.8)$$

For three phase-balanced systems, it can be shown that the transmitted real power is three times the real power per phase:

$$P_{3\Phi} = 3P_{1\Phi} = 3|V_{1\Phi}| \cdot |I| \cdot \cos\phi = \sqrt{3} \cdot |V_L| \cdot |I| \cdot \cos\phi \qquad (A.9)$$

where V_L is the line voltage, which is related to the single-phase voltage:

$$\left|V_L\right| = \sqrt{3} \cdot \left|V_{1\Phi}\right| \tag{A.10}$$

Similar to Equation A1.8, the complex power for three phases is defined as

$$S_{3\Phi} = 3S = 3V \cdot I* \tag{A.11}$$

A.3 PER-UNIT REPRESENTATION

In power system analysis, electrical quantities such as voltages, currents, impedances, and powers are converted in per-unit values. Per-unit system provides a number of favorable attributes such as

1. The range of data obtained is much more limited than the ranges obtained using the real units such as ohms, amperes, and kilovolts. For example, in a power system network analysis, there could be buses with voltage level ranging from 12 kV all the way to 500 kV. While in real units these values are far apart by a factor of about 40, their per-unit values could be very similar. For this reason, the use of actual unit values is computationally less efficient than using per-unit values.

2. Per-unit systems remove the confusion over phase values versus three-phase values and transformer primary and secondary voltages.

In system studies, it is more practical to choose a common base value for voltage (V_b) and MVA (S_b). Assuming that these values are phase quantities, the two values can be used to obtain the base values for phase current (I_b):

$$S_b = V_b \cdot I_b \tag{A.12}$$

The per-unit values for voltage, current, and power can be defined as

$$V_{pu} = \frac{V}{V_b} \tag{A.13}$$

$$I_{pu} = \frac{I}{I_b} \tag{A.14}$$

$$S_{pu} = \frac{S}{S_b} \tag{A.15}$$

In terms of the chosen values, we can define base impedance and base admittance Z_b and Y_b:

$$Z_b = \frac{V_b}{I_b} = \frac{V_b^2}{S_b} \tag{A.16}$$

$$Y_b = \frac{1}{Z_b} = \frac{S_b}{V_b^2} \tag{A.17}$$

and the per-unit values can be defined as

$$Z_{pu} = \frac{Z}{Z_b} = \frac{Z \cdot S_b}{V_b^2} \tag{A.18}$$

$$Y_{pu} = \frac{Y}{Y_b} = \frac{Y \cdot V_b^2}{S_b} \tag{A.19}$$

Sometimes, it is necessary to convert an impedance from one base system to another. Using Equation A1.18, one gets

$$Z_{pu1} = \frac{Z \cdot S_{b1}}{V_{b1}^2} \tag{A.20}$$

$$Z_{pu2} = \frac{Z \cdot S_{b2}}{V_{b2}^2} \tag{A.21}$$

By eliminating Z, the conversion formula is obtained as

$$Z_{pu2} = \frac{Z_{pu1} \cdot S_{b2} \cdot V_{b1}^2}{S_{b1} \cdot V_{b2}^2} \tag{A.22}$$

In the three-phase systems, the base current can be calculated by the following equation:

$$I_{base} = \frac{S_{3ph\,base}}{\sqrt{3}V_{L\,base}} = \frac{S_{3ph\,base}}{3V_{LG\,base}} \tag{A.23}$$

The voltage per-unit base can be selected as single-phase line to ground or line-to-line voltage, where

$$V_{LG\,base} = \frac{V_{L\,base}}{\sqrt{3}} \tag{A.24}$$

The voltage per-unit values on these different per unit bases can be obtained as

$$V_{L\,pu} = \frac{V_L}{V_{L\,base}} \tag{A.25}$$

$$V_{LG\,pu} = \frac{V_{LG}}{V_{LG\,base}} \tag{A.26}$$

Knowing that $V_L = \sqrt{3}V_{LG}$, then

$$V_{LG\,pu} = V_{L\,pu} \tag{A.27}$$

In a similar way, it is shown that

$$S_{1\text{Ph pu}} = S = S_{3\text{ph pu}} \tag{A.28}$$

This is very important because when we say the power is 0.8 per unit then it would not matter whether we discuss single phase or three phase. To derive this important relationship, we divide both sides of Equation A1.11 by $S_{3\text{ph base}}$:

$$S_{3\text{ph pu}} = \frac{3V \cdot I^*}{S_{3\text{ph base}}} = \frac{3V \cdot I^*}{3V_{\text{LG base}} I_{\text{base}}} \tag{A.29}$$

or

$$S_{3\text{ph pu}} = V_{\text{LG pu}} I_{\text{pu}}^* = S_{1\text{ph pu}} \tag{A.30}$$

A.4 MATRIX ALGEBRA

Matrix algebra provides an efficient approach in the formulation and solution of complex engineering problems. This approach provides a disciplined way to manage multivariable and complex problems.

A matrix is defined as a rectangular array of numbers, called elements, arranged in a systematic manner with m rows and n columns. The elements can be real or complex numbers represented as $a_{ij,}$ where i designates the row and j designates the columns as

$$A = \begin{bmatrix} a_{11} & a_{12} & & a_{1n} \\ a_{21} & a_{21} & & a_{2n} \\ & & & \\ a_{n1} & a_{n2} & & a_{nn} \end{bmatrix} \tag{A.31}$$

A few different types of matrices are defined as follows:

Vector: A matrix with a single row or column is called a vector

Transpose of a matrix: If the rows and columns of an $m \times n$ matrix are interchanges, the resultant $n \times m$ matrix is the transpose and is designated by A^t.

Conjugate of a matrix: If all the elements of a matrix are replaced by their conjugates (e.g., replacing the element $a + jb$ by $a - jb$), the resulting matrix is the A conjugate shown as A^*.

Two important matrix operations are

Matrix additions: Matrices of the same dimension can be added in the form of

$A + B = C$, where all elements of $c_{ij} = a_{ij} + b_{ij}$.

Matrix multiplications: Matrix $A(n \times m)$ can be multiplied by $B(m \times q)$ resulting in a matrix $C(n \times q)$ with its elements defined as

$$c_{ij} = \sum_{k=1}^{q} a_{ik} \cdot b_{kj} \tag{A.32}$$

where $i = 1, 2, \ldots m$ and $j = 1, 2, \ldots n$.

A.5 STEADY-STATE COMPONENT MODELING

For power system steady analysis, simple steady-state models are derived for different components such as

1. Transmission lines
2. Transformers and phase shifters
3. Generators
4. Shunts and condensers
5. Loads

The models derived follow a simple π representation given in Figure A.2. The details of each model are briefly described in the sections that follow.

A.5.1 Transmission Lines

Transmission lines are represented by a PI equivalent shown in Figure A.3.

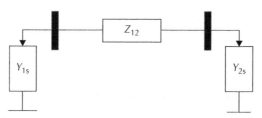

Figure A.2   A typical component steady-state model.

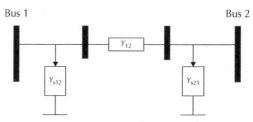

Figure A.3   PI equivalent of a transmission line.

The parameters for the PI equivalent are given by

$$Y_{12} = \frac{1}{Z_{12}} \tag{A.33}$$

$$Z_{12} = \left(\frac{R+JWL}{G+JWL}\right)^2 \sinh(\Psi d) \tag{A.34}$$

$$Y_{s12} = Y_{s21} = \left(\frac{G+JWC}{R+JWL}\right)^{1/2} \tanh\left(\frac{\Psi d}{2}\right) \tag{A.35}$$

where $R+JWL$ is the line impedance per mile; $G+JWC$ is line shunt admittance per mile; d is line length in miles; and Ψ is $[(R+JWL)(G+JWC)]1/2$

A.5.2 Transformers and Phase Shifters

Transformers and phase shifters can be both modeled by using a complex ratio t (magnitude and angle) between the high and low voltage as given in Figure A.4. If the angle is zero, then a transformer is modeled as opposed to a constant magnitude, which represents a phase shifter.

To obtain the transformer PI equivalent similar to the one given in Figure A.1, the values for primary and secondary voltages and currents are related to each other using the following equations:

$$\frac{I_1}{t} = (Y_L + Y_S)V_1 t - Y_L V_2 \tag{A.36}$$

$$I_2 = -t Y_L V_1 + Y_L V_2 \tag{A.37}$$

which gives

$$I_1 = (Y_L + Y_S)t^2 V_1 - Y_L t V_2 \tag{A.38}$$

$$I_2 = -t Y_L V_1 + Y_L V_2 \tag{A.39}$$

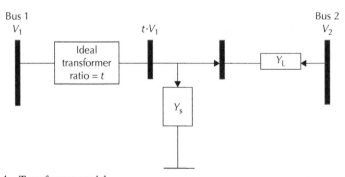

Figure A.4   Transformer model.

Figure A.1, on the other hand, gives the following equations:

$$I_1 = (Y_{S12})V_1 + Y_{L12}(V_1 - V_2) = (Y_{S12} + Y_{L12})V_1 - Y_{L12}V_2 \tag{A.40}$$

$$I_2 = (Y_{S21})V_2 + Y_{L12}(V_2 - V_1) = -(Y_{L12})V_1 + (Y_{S12} + Y_{L12})V_2 \tag{A.41}$$

Comparing these equations with the ones obtained using the PI equivalent of Figure A.1, the following values can be obtained for the PI equivalent:

$$Y_{L12} = Y_L t \tag{A.42}$$

$$(Y_{S12} + Y_{L12}) = (Y_L + Y_S)t^2 \tag{A.43}$$

$$(Y_{S21} + Y_{L12}) = Y_L \tag{A.44}$$

Substituting the value of Y_{L12} from Equation A1.42 into Equations A1.43 and A1.44 gives

$$Y_{S12} = t(t-1)Y_L + t^2 Y_S \tag{A.45}$$

$$Y_{S21} = (1-t)Y_L \tag{A.46}$$

A.5.3 Generators

Only the boundary values of the generator terminal are enough for the generator representation. These include the values of P and specified terminal voltage and the maximum and the minimum values of reactive power, that is, Q_{max} and Q_{min}.

A.5.4 Shunts and Synchronous Condensers

Shunts and condensers produce reactive power. Shunts are represented by their impedances. Synchronous condensers are represented as generators with zero power.

A.5.5 Loads

In steady-state analysis, loads are normally modeled by constant values for P and Q, assuming that they are not voltage-dependent. In more elaborate representations, however, loads can be represented as a function of V.

A.5.6 Network Equations

The network equations are developed by relating voltage and current at each node, giving

$$[I] = [Y][V] \tag{A.47}$$

where I and V are vectors containing voltages and currents at n nodes.

The admittance matrix Y is developed by

Y_{ii} = Sum of all admittances connected to bus i
Y_{ij} = −Admittances between i and j

REFERENCES

1. "Generic System Operator Task List" developed as part of the System Personnel Training Standard PER-005-1, North American Electric Reliability, www.nerc.com, accessed on June 26, 2010.
2. A. Smith, "An Inquiry into the Nature and Causes of the Wealth of Nations," W. Strahan and T. Cadell, London, 1776.
3. "Reliability Standards," North American Electric Reliability Council, www.NERC.com, accessed on November 14, 2013.
4. A. S. Debs, "Modern Power Systems Control and Operation," Kluwer Academic Publishers, Boston, MA, 1988.
5. A. J. Wood and B. F. Wollenburg, "Power Generation Operation and Control," John Wiley & Sons, Inc., New York, 1984.
6. Y. Wallach, "Calculations and Programs for Power System Networks," Prentice Hall, Englewood, NJ, 1984.
7. J. Heydt, "Computer Analysis Methods for Power Systems," Macmillan Publishing Company, New York, 1986.
8. O. Alsac, N. Vempati, B. Stott, and A. Monticelli, "Generalized State Estimation," PICA Conference, Columbus, OH, May 1997.
9. G. Stagg and A. El-Abiad, "Computer Methods in Power System Analysis," McGraw-Hill Book Company, New York, 1968.
10. G. T. Heydt, "Computer Analysis Methods for Power Systems," Macmillan Publishing Company, New York, 1986.
11. B. Stott, "Review of Load Flow Calculation Methods," Proc. IEEE, Vol. 62, pp. 916–929, July 1972.
12. W. Tinnney and J. Walker, "Direct Solutions of Sparse Network Equations by Optimally Ordered Triangular Factorization," Proc. IEEE, Vol. 55, No. 11, pp. 1801–1809, 1967.
13. B. Stott and O. Alsac, " Fast Decoupled Load Flow," IEEE Trans. Power App. Syst., Vol. PAS-96, pp. 273, January/February 1977.
14. P. H. Haley and M. Ayres, "Super Decoupled Load Flow with Distributed Slack," Paper No. 84 WM 046–9 presented at the IEEE Winter Power Meeting, Dallas, TX, January–February 1984.
15. R. A. M. Van Amerongen, "A General Purpose Version of the Fast Decoupled Load Flow," IEEE Trans. Power Syst., Vol. PWRS-4, pp. 760–770, May 1989.
16. W. F. Tinney and A. A. Hart, "Power Flow by Newton's Method," IEEE Trans. Power App. Syst., Vol. PAS, Vol. 21, No. 4, pp. 1550–1556, November 1967.
17. W. F. Tinney, V. Brandwajn, and S. M. Chan, "Sparse Vector Methods," IEEE Trans. Power App. Syst., Vol. PAS-104, No. 2, pp. 295–301, February 1985.

Practical Power System Operation, First Edition. Ebrahim Vaahedi.
© 2014 The Institute of Electrical and Electronics Engineers, Inc. Published 2014 by John Wiley & Sons, Inc.

18. M. Lotfalian, R. Schlueter, D. Idizior, P. Rusche, S. Tedeschi, L. Shu, A. Yazdankhah, "Inertial, Governor and AGC/Economic Dispatch Loadflow," IEEE Power Eng. Rev., Vol. PER-5, No. 11, p. 26, 1985.

19. W. Y. Ng, "Generalized Generation Distribution Factors for Power System Security Evaluations," IEEE Trans. Power App. Syst., Vol. PAS-100, pp. 1001–1005, March 1981.

20. B. Stott, O. Alsac, and A.J. Monticelli, "Security Analysis and Optimization," Proceedings of IEEE, Vol. 5, No. 12, December 1987.

21. O. Alsac, B. Stott, and W. F. Tinney, "Sparsity-Oriented Compensation Methods for Modified Network Solutions," IEEE Trans. Power App. Syst., Vol. PAS-102, pp. 1050–1060, May 1983.

22. J. Zaborszky, K. W. Whang, and K. Prasad, "Fast Contingency Evaluation Using Concentric Relaxation," IEEE Trans. Power App. Syst., Vol. PAS-99, pp. 28–36, January/February 1980.

23. V. Brandwajn and M. G. Lauby, "Complete Bounding Method for AC Contingency Screening," IEEE Trans. Power Syst., Vol. PAS-4, No. 2, pp. 724–729, May 1989.

24. F. F. Wu and A. Monticelli, "Critical Review of External Network Modelling for On-line Security Analysis," Int. J. Electr. Power Energ. Syst., Vol. 5, pp. 222–235, October 1983.

25. W. F. Tinney and J. M. Bright, "Adaptive Reduction for Power Flow Equivalents," IEEE Trans. Power Syst., Vol. PWRS-2, pp. 351–360, May 1987.

26. Y. Chen and A. Bose, "Adaptive Pre-Filter for Voltage Contingency Selection Function," IEEE Trans. Power Syst., Vol. PWRS-5, No. 4, pp. 1478–1656, November 1990.

27. T. F. Haplin, R. Fischl, and R. Fink, "Analysis of Automatic Contingency Selection Algorithms," IEEE Trans. Power App. Syst., Vol. PAS-103, pp. 938–945, May 1984.

28. G. D. Irisarri and A. M. Sasson, "An Automatic Contingency Selection Method for On-line Security Analysis," IEEE Trans. Power App. Syst., Vol. PAS-100, pp. 1838–1844, April 1981.

29. O. E. Elgerd, "Electric Energy Systems Theory: An Introduction," McGraw-Hill Book Company, New York, 1982.

30. C. A. Gross, "Power System Analysis," John Wiley & Sons, Inc., New York, 1979.

31. P. Kundur, "Power System Stability and Control," McGraw-Hill Book Company, New York, 1994.

32. A. Shalaby, V. F. Carvalho, and J. A. Findlay, "On-line Computer Monitoring of Complex System Limits," PICA Conference, IEEE, Cleveland, Ohio, pp. 64–72, May 1979.

33. T. Van Cutsem and C. Vournas, "Voltage Stability of Electric Power System," Kluwer Academic Publishing, Boston, MA, 1998.

34. IEEE/PES, C. Canizares ,"Voltage Stability Assessment: Concepts, Practices and Tools," Technical Report, Special Publication of IEEE Power System Stability Subcommittee, August 2002.

35. CIGRE TF 38.02.12, "Criteria and Countermeasures for Voltage Collapse," Final Report, CIGRE, Paris, December 1994.

36. WSCC RRWG, "Proposed Voltage Stability Guidelines, Under Voltage Load Shedding Strategy, and Reactive Power Reserve Monitoring Methodology," Final Report, September 1997.

37. V. Ajjarapu and C. Christ, "Continuation Power Flow; a Tool for Steady State Voltage Stability Analysis," IEEE Trans. Power Syst., Vol. PWRS-7, No. 1, pp. 416–423, February 1992.

38. T. Van Cutsem and R. Mailhot, "Validation of a Fast Voltage Stability Analysis Method on the Hydro-Québec System," IEEE Trans. Power Syst., Vol. PWRS-12, pp. 282–292, 1997.

39. E. Vaahedi, C. Fuchs, W. Xu, Y. Mansour, H. Hamdanizadeh, and G. K. Morison, "Voltage Stability Contingency Screening and Ranking," IEEE Tran. Power Syst., Vol. 14, No. 1, pp. 256–265, February 1999.

40. NERC Automatic Generation Control Standards, BAL-005-0b, North American Electric Reliability Council, May 2, 2007.
41. WECC Field Trial of the Balancing Authority ACE Limit Under NERC Project 2007–18 Reliability-Based Control, Western Electric Coordinating Council, Draft 9, January 27, 2010.
42. D. N. Ewart, "Automatic Generation Control – Performance Under Normal Conditions" Systems Engineering for Power: Status and Prospects, US. Government Document CONF-750867, pp. 1–14, 1975.
43. N. Jaleeli, L. S. VanSlyck, D. Edwart, L. H. Fink, and A. G. Hoffman, "Understanding Automatic Generation Control," IEEE Trans. Power Syst., Vol. PWRS-7, No. 3, pp. 1106, August 1992.
44. H. Glavitsch, "Discussion of Reference 2," IEEE Trans. Power Syst., Vol. PWRS-7, No. 3, p. 1106, August 1992.
45. M. Fekri Moghadam, M. Metcalf, W. G. Dunford, and M. Metcalf, "Using Industrial Load Flexibility to Increase Hydroelectric Generation Efficiency," Paper Submitted for Publication to IEEE Transactions in Power Systems.
46. X.-P. Zhang, "Restructured Electric Power Systems: Analysis of Electricity Markets with Equilibrium Models," John Wiley & Sons, Inc., Hoboken, NJ, p. 330, July 2010,
47. R. B Wilson, "Architecture of Power Markets," Econometrica, Vol. 70, pp. 1299–1340, July 2002.
48. J. H. Chow, R. W. De Mello, and K. W. Cheung, "Electricity Market Design: An Integrated Approach to Reliability Assurance," IEEE Trans. Power Syst., Vol. PWRS-93, No. 11, pp. 1956–1969, November 2005.
49. J. Carpentier, "Contribution a l'etude du dispatching economique," Bull. Soc. Fr., pp. 431–447, 1962.
50. H. W. Dommel and W. F. Tinney, "Optimal Power Flow Solutions," IEEE Trans. Power App. Syst., Vol. PAS-97, pp. 37–47, 1968.
51. E. Vaahedi, Y. Mansour, J. Tamby, W. Li, W., and D. Sun, "Evaluation of Existing OPF/Var Planning Methods on Four Large Scale Utility Systems," IEE P-Gener. Transm. D., July 2000.
52. H. Kuhn, and A. W. Tucker, "Nonlinear Programming." Proceedings of 2nd Berkeley Symposium, University of California Press, Berkeley, CA, pp. 481–492. http://project euclid.org/euclid.bsmsp/1200500249. MR47303, accessed on November 14, 2013.
53. W. Karush, "Minima of Functions of Several Variables with Inequalities as Side Constraints," M.Sc. Dissertation, Dept. of Mathematics, Univ. of Chicago, Chicago, IL, 1939.
54. J. Carpentier, "Differential Injection Method, a General Method for Secure and Optimal Load Flows," PICA, Minneapolis, MN, pp. 255–262, 1973.
55. R. C. Burchett, H. H. Happ, and D. R. Vierath, "Quadratically Convergent Optimal Power Flow," IEEE Trans. Power App. Syst., Vol. PAS-103, pp. 3267–3275.
56. D. I. Sun, B. Ashley, B. Brewer, A. Hughes, and W. F. Tinney, "Optimal Power Flow by Newton Approach," IEEE Trans. Power App. Syst., Vol. PAS-103, pp. 2864–2880, October 1984.
57. G. A. Maria and J. A. Findlay, "A Newton Optimal Power Flow for Ontario Hydro EMS," IEEE Trans. Power Syst., Vol. PWRS-2, pp. 576–584, August 1987.
58. O. Alsac, J. Bright, M. Prais, and B. Stott, "Further Development in LP-Based Optimal Power Flow," IEEE Trans. Power Syst., Vol. 5, No. 3, pp. 697–711, August 1990.
59. A. Monticelli and W. E. Liu, "Adaptive Movement Penalty Method for the Newton Optimal Power Flow," IEEE Trans. Power Syst., Vol. PWRS-7, pp. 334–342, November 1988.
60. N. Karmarker, "A New Polynomial Time Algorithm for Linear Programming," Combinatorica, Vol. 4, pp. 373–395, 1984.
61. C. N. Lu and M. R. Unum, "Network Constrained Security Control Using an Interior Point Algorithm," IEEE Trans. Power Syst., Vol. PWRS-8, pp. 1068–1076, August 1993.

62. S. Granville, "Optimal Reactive Dispatch through Interior Point Methods," IEEE Trans. Power Syst., Vol. PWRS-9, pp. 136–146, February 1994.

63. K. Clements, "Interior-Point Optimization Methods," in Advanced Optimization Techniques, PICA Tutorial, Baltimore, MD, May 1991.

64. Y. Hong, D. I. Sun, S. Lin, and C. Lin, "Multi-Year Multi-Case Optimal Var Planning," IEEE Trans. Power Syst., Vol. PWRS-5, pp. 1294–1301, November 1990.

65. S. Granville, M. V. F. Pereira, and A. Monticelli, "An Integrated Method for Var Sources Planning," IEEE Trans. Power Syst., Vol. PWRS-3, pp. 1741–1747, May 1988.

66. E. Vaahedi and M. H. Zein Eldin, "Considerations in Applying Optimal Power Flow to Power System Operation," IEEE Trans. Power Syst., Vol. PWRS-4, pp. 84–89, May 1989.

67. E. Vaahedi, Y. Mansour, J. Tamby, W. Li, and D. Sun, "Large Scale Voltage Stability Constrained Var Planning and Voltage Stability Application Using Existing OPF/Optimal Var Planning Tools," IEEE Trans. Power Syst., Vol. PWRS-14, pp. 65–74, February 1999.

68. A. J. Monticelli, M. V. F. Pereira, and S. Granville, "Security Constrained Optimal Power Flow with Post-contingency Corrective Rescheduling," IEEE Trans. Power Syst., Vol. PWRS-2, pp. 175–182, February 1987.

69. V. Vankayala, E. Vaahedi, D. Cave, and M. Huang, "Opening Up for Interoperability—A Pragmatic Open Architecture for a Control Center Consolidation Project," IEEE Power and Energy Magazine, March/April 2008.

70. T. E. Dy-Liacco, "Real-time Computer Control of Power Systems Modern Control Centers and Computer Networking," Proceedings of IEEE Computer Applications in Power, July 1974.

71. E. Vaahedi, E. A. Chang, N. Muller, S. Mokhtari, and G. Irrisari, "A Future Application Environment for B.C. Hydro's EMS," IEEE Trans. Power Syst., Vol. PWRS-16, pp. 9–14, February 2001.

72. A. Handschin and E. Petroianu, "Energy Management Systems, Operations and Control of Electric Energy Transmission Systems," Springer-Verlag, Berlin, Heidelberg, 1991.

73. T. E. Dy-Liacco, "Modern Control Centers and Computer Networking," IEEE Comput. Appl. Power, Vol. 7, pp. 17–22, 1994.

74. E. Vaahedi and M. Shahidehpour, "Decision Support Tool Requirement in Restructured Energy Systems," IEEE Trans. Power Syst., May 2004.

75. J. W. Evans, "Energy Management System, Survey of Architecture," IEEE Comput. Appl. Power, Vol. 2, pp. 11–16, January 1989.

76. R. Kalisch , "EMS for the 21st Century," CIGRE Working Group D2.24, February 2011.

77. R. W. Uluski, "Interactions between AMI and Distribution Management Systems for Efficiency/Reliability Improvements at a Typical Utility," Power and Energy Society General Meeting—Conversion and Delivery of Electrical Energy in the 21st Century, IEEE, July 20–24, 2008.

78. R. Langsdon, "DMS—The Integration Solution for GIS/SCADA/OMS," GITA's 23rd Annual Conference and Exhibition, Denver, CO, March 26–29, 2000.

79. Review of On-line Dynamic Security Assessment Tools and Techniques Working Group 601 of Study Committee C4, Final Report, CIGRE, Paris, January 2007.

80. J. Viikinsalo, A. Martin, K. Morison, L. Wang, and F. Howell, "Transient Security Assessment in Real-Time at Southern Company," Paper presented at the IEEE PSC&E Conference, IEEE, Atlanta, GA, Oct 29–November 1, 2006.

81. I. M. Dudurych, "On-line Assessment of Secure Level of Wind on the Irish Power System," Paper accepted for presentation at the IEEE 2010 Summer Meeting, Minneapolis, MN, 2010.

82. H. D. Chiang, "The BCU Method for Direct Stability Analysis of Electric Power Systems: pp. Theory and Applications" Systems Control Theory for Power Systems, Volume 64 of IMA Volumes in Mathematics and its Applications, pp. 39–94. Springer-Verlag, New York, 1995.

83. H.-D. Chiang, J. Tong, and Y. Tada, "On-line Transient Stability Screening of 14,000-Bus Models Using TEPCO-BCU," Paper accepted for presentation at the IEEE 2010 Summer Meeting, Minneapolis, MN, 2010.

84. M. Pavella, D. Ernst, and D. Ruiz-Vega, "Transient Stability of Power Systems: A Unified Approach to Assessment and Control," Kluwer Academic Publishers, Boston, MA, 2000.

85. Y. Mansour, E. Vaahedi, and M. A. El-Sharkawi, "Dynamic Security Contingency Screening and Ranking Using Neural Networks," IEEE Trans. Neural Networks, Vol. 8, No. 4, pp. 942–950, July 1997.

86. E. Vaahedi, "Assessment of Voltage Security Methods and Tools," EPRI Report TR-105214, Final Report, prepared by B.C. Hydro, October 1995.

87. North American Electric Reliability Standard MOD-001-0—Documentation of TTC and ATC Calculation Methodologies, April 1, 2005.

88. North American Electric Reliability Standard MOD-008-1—Transmission Reliability Margin Calculation Methodology, August 26, 2008.

89. W. Li, E. Vaahedi, and Z. Lin, "BC Hydro's Transmission Reliability Margin Assessment in Total Transfer Capability Calculations," IEEE Trans. Power Syst., Vol.28, pp. 4796–4802, November 2013.

90. M. Patel , "Real-Time Application of Synchrophasors for Improving Reliability," Paper prepared by NERC's Real-Time Application of Synchrophasors for Improving Reliability Task Force, October 18, 2010.

91. J. Giri, M. Parashar, J. Trehern, and V. Madani, "The Situation Room – Control Center Analytics for Enhanced Situational Awareness," IEEE Power Energy Mag., Vol. 10, pp. 24–39, September/October 2012.

92. Synchrophasor System Benefits Fact Sheet, North American Synchrophasor Initiative (NASPI), March 2009, https://www.naspi.org/, accessed on November 14, 2013.

93. Z. Sumic, "Hype Cycle for Smart Grid Technologies, 2012," Gartner Research Note, July 27, 2012.

94. Distributed Energy Resource Basics, US Department of Energy, March 2, 2012, http://www1.eere.energy.gov/femp/technologies/derchp_derbasics.html, accessed on November 14, 2013.

95. U. A. Dougiass and A. Edris, "Real-Time Monitoring and Dynamic Thermal Rating of Power Transmission Circuits," IEEE Trans. Power Deliv., Vol. 10, No. 3, pp. 1460–1470, July 1995.

96. Report to Congress Prepared by staff of the Federal Energy Regulatory Commission and the U.S. Department of Energy, "Implementation Proposal for The National Action Plan on Demand Response," July 2011. http://energy.gov/sites/prod/files/oeprod/Documents andMedia/ImplementationProposalforNAPDRFinal.pdf, accessed on November 25, 2013.

97. M. F. Moghadam, M. Metcalfe, E. Vaahedi, and W. G. Dunford, "Application of Industrial Demand Management in Economic Load Dispatch," Paper accepted for presentation at the IEEE PES General Meeting, Vancouver, BC, Canada, July 2013.

98. North American Electric Reliability Corporation (NERC) Report, "Real-Time Tools Survey Analysis and recommendation," Paper prepared by NERC's Real-Time Tools Best Practices Task Force, NERC, March 13, 2008.

99. E. Vaahedi, W. Li, T. Chia, and H. Dommel, "Large Scale Probabilistic Transient Stability Assessment Using B.C. Hydro's On-line Tools," IEEE Trans. Power Syst., Vol. PWRS-15, No. 2, pp. 661–667, May 2000.

INDEX

Practical Power System Operation, First Edition. Ebrahim Vaahedi.
© 2014 The Institute of Electrical and Electronics Engineers, Inc. Published 2014 by John Wiley & Sons, Inc.

IEEE Press Series on Power Engineering

Series Editor: M. E. El-Hawary, Dalhousie University, Halifax, Nova Scotia, Canada

The mission of IEEE Press Series on Power Engineering is to publish leading-edge books that cover the broad spectrum of current and forward-looking technologies in this fast-moving area. The series attracts highly acclaimed authors from industry/academia to provide accessible coverage of current and emerging topics in power engineering and allied fields. Our target audience includes the power engineering professional who is interested in enhancing their knowledge and perspective in their areas of interest.

Printed and bound by CPI Group (UK) Ltd, Croydon, CR0 4YY

16/04/2025

14658592-0004